T0337887

BOOLEAN CIRCUIT REWIRING

BOOLEAN CIRCUIT REWIRING:

BRIDGING LOGICAL AND PHYSICAL DESIGNS

Tak-Kei Lam

The Chinese University of Hong Kong, Hong Kong, P. R. China

Wai-Chung Tang

Queen Mary University of London, UK

Xing Wei

Easy-Logic Technology Ltd. Hong Kong, Hong Kong, P. R. China

Yi Diao

Easy-Logic Technology Ltd. Hong Kong, Hong Kong, P. R. China

David Yu-Liang Wu

Easy-Logic Technology Ltd. Hong Kong, Hong Kong, P. R. China

This edition first published 2016

Registered office
John Wiley & Sons Singapore Pte Ltd, 1 Fusionopolis Walk, #07-01 Solaris South Tower, Singapore 138628.

For details of our global editorial offices, for customer services and for information about how to apply for permission to reuse the copyright material in this book please see our website at www.wiley.com.

Library of Congress Cataloging-in-Publication Data applied for.

ISBN: 9781118750117

A catalogue record for this book is available from the British Library.

Set in 10/12pt, TimesLTStd by SPi Global, Chennai, India.

Printed and bound in Singapore by Markono Print Media Pte Ltd

1 2016

Contents

List of Figures

List of Tables

Preface

Our group has been working on wire-based logic restructuring (rewiring) for over a decade. Over the years, we have published numerous conference and journal papers on rewiring. As a recent major milestone, we have developed a rewiring scheme that reaches a near-complete rewiring rate (96%). This result demonstrates the high power of this kind of logic transformation techniques and the great potential of applying them on modern electronic design automation (EDA) tools.

Because of the aggressive and continuous scaling down of transistor sizes, to 45, 22 nm, and even below 16 nm, wires have become a dominant factor affecting circuit performance. Hence, rewiring is particularly suitable for today's nanometer technologies.

We could not find a book that focuses on and discusses rewiring techniques. Since rewiring techniques have become much more practical in nanometer technologies, we felt there was a need to publish a reference book to provide readers with the key ideas.

This book is of introductory to intermediate level. We hope this book will help in popularizing science, and the readers will find this book interesting and informative.

TAK-KEI LAM, WAI-CHUNG TANG, XING WEI,
YI DIAO, AND DAVID YU-LIANG WU

Introduction

The concepts of various major rewiring techniques are explained throughout the book gradually. First, readers will be presented with the basic ideas of rewiring. Next, the technical details of each kind of rewiring technique will be discussed in detail. Finally, the applications of rewiring techniques in various electronic design automation (EDA) areas will be introduced.

Intended Audience

Students studying computer/electronic engineering, academic staff, and even EDA engineers are the intended readers of this book. The readers should have some basic knowledge of Boolean algebra, logic gates, and graph theory. For readers without the related advanced knowledge, essential concepts will be introduced and explained throughout the book.

Type Conventions

The following conventions are used in this book:

- Mathematical symbols and names of circuit elements, such as a and α, are typeset in this font.
- `Codes are typeset in this font.`

Acknowledgments

Many people have contributed to this book in the forms of research output, implementations of algorithms, suggestions for content, and, last but not least, simply being encouraging. This book could never have been completed without their generous effort. We are very grateful to the following people for all they have done:

- The authors of RAMBO, REWIRE, RAMFIRE, GBAW, IRRA, NAR, ECR, FECR, CECR, SPFD-based and all other rewiring techniques.
- The authors of the typesetting system LaTeX and the plugins.

1

Preliminaries

1.1 Boolean Circuits

A Boolean variable is a variable whose value can only be either 0 (false) or 1 (true) or unknown. Every Boolean variable has two literals. They are the normal form and the negation/complement of the variable. The negation of a variable always evaluates to the opposite value of the variable. Suppose v is a Boolean variable; then its negation is $\bar{v}$. When v is 1, $\bar{v}$ is 0; when v is 0, $\bar{v}$ is 1. The literals of variable v are then v and $\bar{v}$.

A function consisting of Boolean variables is known as a Boolean function. It is a mapping between Boolean spaces. For example, the function $f : B^m \rightarrow B^n$ is a mapping between the input space of m Boolean variables and the output space of n Boolean variables. We use $f(x_1, x_2, \ldots, x_{m-1}, x_m)$ to indicate the input variables or input values of the Boolean function f.

The mapping between Boolean spaces is achieved by Boolean operators. The basic Boolean operators (operations) AND, OR, NOT, XOR, and XNOR are denoted as $\cdot$, $+$, $\sim$, $\oplus$, and $\bar{\oplus}$, respectively, in this book. We may omit the symbol $\cdot$ for clarity. The behavior of the basic operators is listed in Table 1.1. Complex Boolean operators can be derived from these basic operators. In fact, only AND and NOT, or only OR and NOT, are sufficient to derive all other Boolean operations.

An example of Boolean function is $f(a, b) = a \cdot b$, which computes the logical conjunction of variables a and b. A Boolean function may contain literals. The Boolean function $f(a, b) = a \cdot \bar{b}$ is such an example that computes the logical conjunction of variable a and the negation of variable b. It may be surprising for readers who are not familiar with Boolean algebra to see a function $f(a, b, c) = a \cdot b$. This function is actually nothing special but is normal and valid. It just means that, among the three variables, the value of variable c is "don't care." That is to say, the value of c can be either 0 or 1, and $f(a, b, c) = ab = ab\bar{c} + abc$. For another example, the function $f(a, b, c) = (a + b)$ can be expanded into $f(a, b, c) = (a + b) = (a + b)\bar{c} + (a + b)c$.

Observability don't cares (ODCs) (Damiani and De Micheli 1990) of a Boolean variable are the conditions under which the variable is not affecting any of the primary outputs. For example, if an input i of an AND gate has the controlling value 0, its set of other inputs J are unobservable no matter what values they have. The ODC of J is $\bar{i}$. Satisfiability don't cares

Boolean Circuit Rewiring: Bridging Logical and Physical Designs, First Edition.
Tak-Kei Lam, Wai-Chung Tang, Xing Wei, Yi Diao and David Yu-Liang Wu.

Table 1.1 Behavior of the basic Boolean operators

Operator	When will it returns true?
AND $\cdot$	All of its operands are true
OR $+$	Any one of its operands is true
NOT $\sim$	Its operand is false
XOR $\oplus$	Both of its operands have different values
XNOR $\bar{\oplus}$	Both of its operands have the same values

(SDCs) of a circuit node represent the local input patterns at the node that cannot be generated by the node's fanins. As a trivial example of SDC, if we connect all inputs of a two-input AND gate to a common signal, the values of its inputs can never be $\{1, 0\}$ or $\{0, 1\}$.

Many rules in ordinary algebra, such as commutative addition and multiplication, associative addition and multiplication, and variable distribution, can be applied into Boolean algebra. Therefore, function $f(a, b, c) = (a + b) = (a + b)\bar{c} + (a + b)c = a\bar{c} + b\bar{c} + ac + bc$. For each of the conjunction term, it can be expanded by connecting it with all combinations of the literals of the missing variables by conjunction. Some additional important rules that are obeyed in Boolean algebra only include $a \cdot a = a$ and $a + a = a$. Regarding our example, it can be expanded as follows:

$$\begin{aligned} f(a, b, c) &= (a + b) \\ &= (a + b)\bar{c} + (a + b)c \\ &= (a\bar{c} + b\bar{c}) + (ac + bc) \\ &= (ab\bar{c} + a\bar{b}\bar{c} + ab\bar{c} + \bar{a}b\bar{c}) + (abc + a\bar{b}c + abc + \bar{a}bc) \\ &= ab\bar{c} + a\bar{b}\bar{c} + \bar{a}b\bar{c} + abc + a\bar{b}c + \bar{a}bc \end{aligned}$$

Other rules can be derived from the basic rules easily. Since Boolean algebra is a vast area of study, even the elementary topics can cover a whole book. In this book, we shall not cover every detail.

Boolean functions can be realized in hardware using logic gates. A Boolean circuit or Boolean network is composed of gates and implements some Boolean functions. We simply use circuits or networks to represent Boolean circuits when the meaning is clear in the context. Figure 1.1 illustrates the logic gates implementing the basic Boolean functions. An example circuit composed of AND, OR, and NOT gates is shown in Figure 1.2(b). For gate e, its inputs are a and b, so it is implementing the function $a + b$. Regarding gate f, its function is $a \cdot \bar{b}$.

A less famous Boolean operator is the cofactor. The cofactor of a Boolean function $f(x_1, x_2, \dots, x_{n-1}, x_n)$ with respect to a variable x_k is $f|_{x_k} = f(x_1, x_2, \dots, x_{k-1}, 1, x_{k+1}, \dots, x_{n-1}, x_n)$. Suppose $f = ab + c$, $f|_a = b + c$, and $f|_c = ab + 1 = 1$. Similarly, the cofactor of a Boolean function $f(x_1, x_2, \dots, x_{n-1}, x_n)$ with respect to the complement of a variable x_k is $f|_{\overline{x_k}} = f(x_1, x_2, \dots, x_{k-1}, 0, x_{k+1}, \dots, x_{n-1}, x_n)$. Suppose $f = ab + c$, $f|_{\bar{a}} = 0 \cdot b + c = c$. Every Boolean function can be expressed using Shannon's expansion. For example, $f(x) = x \cdot f|_x + \bar{x} \cdot f|_{\bar{x}}$. An example of a function with multiple inputs is $f(x, y, z, w) = xyf|_{xy} + \bar{x}yf|_{\bar{x}y} + x\bar{y}f|_{x\bar{y}} + \bar{x}\bar{y}f|_{\bar{x}\bar{y}}$

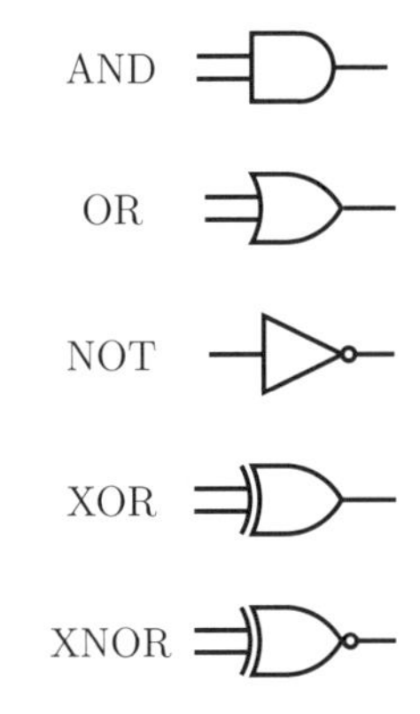

Figure 1.1 Basic logic gates

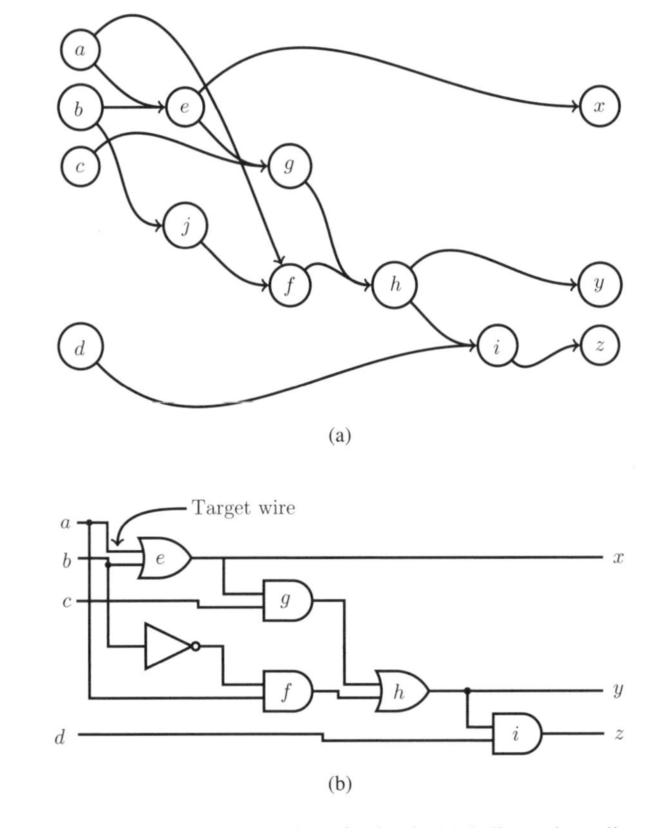

Figure 1.2 Directed acyclic graph representation of a circuit. (a) A directed acyclic graph; (b) a Boolean circuit

In general, Boolean circuits can be represented by directed acyclic graphs (DAGs). A DAG is a kind of graph in which two vertices may be connected by an edge pointing to either one of the vertices, such that there are no loops in the graph. With regard to a circuit, the vertices of its corresponding graph represent its gates, and the edges of the graph represent its wires. An example of a DAG is illustrated in Figure 1.2(a). It is in fact the corresponding graph representation for the circuit in Figure 1.2(b). For instance, the node e in the graph represents the OR gate e, and the edge from a to e corresponds to the target wire indicated in the figure.

Each node in a DAG is therefore associated with a Boolean function. Edges pointing toward a node are known as the fanins of the node, and edges leaving the node are known as the fanouts of the node. The source of a wire connecting s to d has a fanout d, and the destination of the wire d has a fanin s. The number of fanins of a node n is called the in-degree/fanin number of the node, which is denoted by $d^{+}(n)$. Similarly, the number of fanouts of node n is called the out-degree/fanout number and is denoted by $d^{-}(n)$. A special case of a DAG is the and-inverter graph (AIG) in which every node represents an AND gate. The values of the edges may be complemented to implement all Boolean functions.

In a circuit, the nodes whose fanin numbers are 0 or have no fanins are known as primary inputs (PIs), and the nodes whose fanout numbers are 0 or have no fanouts are known as primary outputs (POs).

Node a is a transitive fanin (TFI) of node b if there is a path from node a to node b. Node b is said to be a transitive fanout (TFO) of node a if is reachable from node a. All TFIs of a node form the TFI cone of the node, and all TFOs of a node form the TFO cone of the node.

The most famous problem regarding Boolean circuits is the Boolean satisfiability problem. It is always known as the SAT problem. This problem is to determine whether there is any assignment of values to the variables of a given Boolean function such that the function can be evaluated to 1. For the Boolean function $f(a, b, c) = a \cdot b + c$, the value assignments $\{a = 1, b = 1, c = 0\}$ allows f to evaluate to 1. It is a solution that satisfies this SAT problem instance.

1.2 Redundancy and Stuck-at Faults

Central to rewiring techniques is the concept of logic redundancy. Logic redundancy means that, in a circuit, the logic value of a connection or a component has no effect on the output values of the circuit. The occurrence of logic redundancy is usually due to some design errors or faults such as manual design errors, shorting of two wires, or fabrication defects in the circuit. In practice, this kind of redundancy is always targeted for removal so as to minimize the chip area and to improve the yield of the chip. However, there are other kinds of redundancy that are added into circuits intentionally, for example, for fault tolerance.

Engineers have developed the stuck-at fault model to manipulate logic redundancy. As a matter of fact, many types of faults in Boolean circuits can be modeled as stuck-at faults (Jha and Kundu 1990). In this model, the value of a faulty wire is stuck at a constant, either 0 ($sa0$) or 1 ($sa1$). The input vectors that can test a fault are the test vectors of the fault. If there is no test vector for a fault, the fault is regarded as untestable. Testing an untestable fault always results in value assignment conflicts at some locations. Regarding an AND (OR) gate input wire, it is sufficient to certify the untestability of its stuck-at-1 (stuck-at-0) fault to ensure that it is redundant. Its stuck-at-1 (stuck-at-0) is called a noncontrolling-value stuck-at fault, otherwise called a controlling-value stuck-at fault. The names come from the fact that the

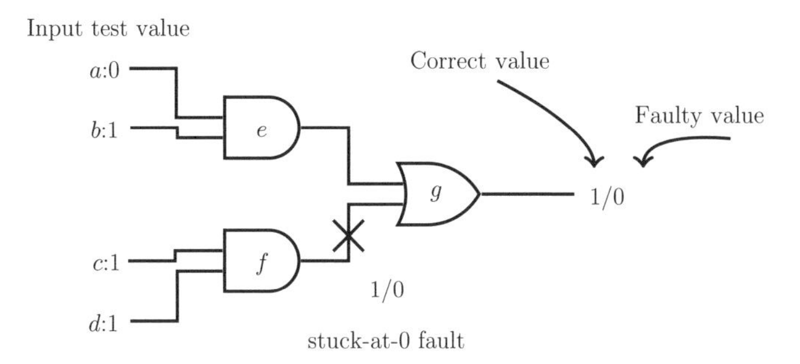

Figure 1.3 Stuck-at fault

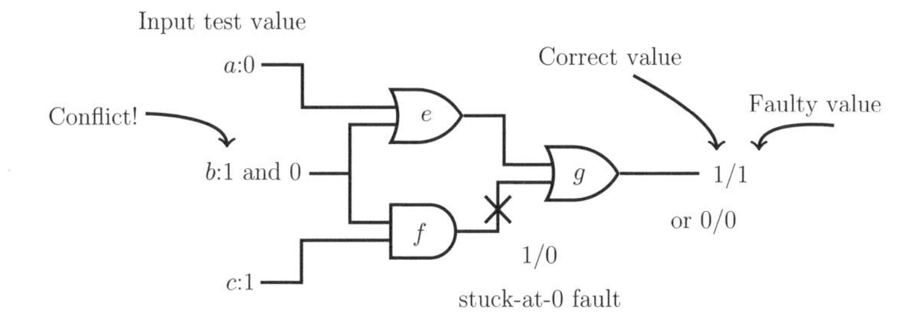

Figure 1.4 Untestable stuck-at fault

controlling value of an AND (OR) gate is 0 (1). In general, a gate's input wire is redundant if its noncontrolling-value stuck-at fault cannot be tested. For an input of a gate that has no noncontrolling value (e.g., XOR), it is redundant only if both its noncontrolling-value and controlling-value stuck-at faults are untestable.

In this book, thc notation $sa0(s \rightarrow d)$ $(sa1(s \rightarrow d))$ will bc uscd to dcnotc thc stuck-at-0 (stuck-at-1) of the wire $s \rightarrow d$ whose source node (source) is s and destination node (sink) is d. The D-notation is also adopted to denote stuck-at-0 as $1/0(D)$ and stuck-at-1 as $0/1(\bar{D})$.

Figure 1.3 shows a stuck-at fault. The notation $1/0$ on the wire means that the normal or correct value of the wire is 1 and is 0 when a fault occurs. To observe such a fault, we need to find a path in the circuit so that the faulty value can be propagated to at least one primary output and the difference between the normal and faulty values can be observed by the users. The value assignment $\{a = 0, b = 1, c = 1, d = 1\}$ is called a test vector for propagating and observing the $1/0$ fault in the figure. If such a path does not exist for a stuck-at fault, then we can say that there is no test vector for the fault and the fault is undetectable. The wire is thus redundant and can be removed without changing the behavior of the circuit.

An untestable stuck-at fault is shown in Figure 1.4. The value of the input b has to be set to 1 because we want to assign 1 as the correct value of the stuck-at-0 fault. On the other hand, it has to be set to 0 simultaneously so as to make the fault observable. This is obviously impossible. If b is either 1 or 0, the correct and faulty values at the circuit's output are either $1/1$ or $0/0$. This means the stuck-at-0 fault cannot be tested at all. From this analysis, we know that the corresponding wire $f \rightarrow g$ is redundant.

1.3 Automatic Test Pattern Generation (ATPG)

Automatic Test Pattern Generation (ATPG), as its name implies, is a circuit testing method and is commonly applied to prove the testability of the stuck-at faults in a given circuit. With regard to a stuck-at fault, it works by feeding the circuit with the essential input test vectors and comparing the output values obtained with those of a fault-free circuit. If there is any difference, the fault is observable.

There are two steps in ATPG, namely (i) fault excitation (activation) and (ii) fault propagation. Fault excitation for a wire's stuck-at fault is to derive a test vector such that the wire can have distinguishable correct and faulty values. Regarding the stuck-at-0 fault in Figure 1.3, the output of gate f has to be 1 to differentiate the correct and faulty values at wire $f \rightarrow g$. Then, the values of both c and d have to be 1 because an AND gate can only output 1 when all of its inputs are 1. The partial test vector is $\{c = 1, d = 1\}$.

Fault propagation means propagating a fault to some outputs of the circuit. If the upper input of the OR gate g (wire $e \rightarrow g$) is assigned with 1, the circuit's output (g's output) will be $1/1$ and the fault is not observable. Hence, wire $e \rightarrow g$ should be assigned with 0 instead. The value of wire $e \rightarrow g$ will be 0 when one of the inputs of gate e is 0. Therefore, all the test vectors for the stuck-at-0 fault in this example are as follows: (i) $\{a = 0, b = 0, c = 1, d = 1\}$ (ii) $\{a = 0, b = 1, c = 1, d = 1\}$ (iii) $\{a = 1, b = 0, c = 1, d = 1\}$. Alternatively, the test vectors can be represented by (i) $\{a = *, b = 0, c = 1, d = 1\}$ (ii) $\{a = 0, b = *, c = 1, d = 1\}$, where $*$ means that the value can be either 0 or 1.

1.4 Dominators

A dominator (Kirkland and Mercer 1987) of a wire w is a gate through which all paths from w to any primary outputs must go. For instance, in Figure 1.4, the set of dominators of b is g because the paths from b to all the circuit outputs (i) $b \rightarrow e \rightarrow g \rightarrow$ (ii) $b \rightarrow f \rightarrow g \rightarrow$ all pass through g. Side inputs of a dominator with respect to a wire w are its inputs that are not lying in w's fault propagation paths. In Figure 1.4, the upper input ($e \rightarrow g$) of the dominator g with respect to the stuck-at fault $sa0(f \rightarrow g)$ is a side input. The gates e and f are not the dominators of $sa0(f \rightarrow g)$.

Algorithm 1.1: Procedure *Cal_Dominators*

```
  input : a circuit C
1 begin
2   N ← primary inputs of C;
3   foreach node n ∈ N do
4     D_n ← n;                /* D_n is the set of dominators of n */
5     S ← ∅;
6     foreach fanout f_n of n do
7       S ← S ∩ Cal_Dominators(f_n);
8     D_n ← D_n ∪ S;
9 end
```

By definition, a gate is a dominator of itself because it must pass through itself. Every fanout of the gate has its own set of dominators. The common set of the dominators of the gate's fanouts represents the gates through which all fanouts must pass. Hence, the dominators of all the gates in a circuit can be calculated recursively. The recursive algorithm is listed in Algorithm 1.1.

For a given fault, determining all test vectors to detect the fault is proven to be extremely complex. The notion of dominators is always applied in fault propagation so that the process is more systematic and efficient. In fault propagation, the side inputs of each of the dominators with respect to a fault are assigned with the noncontrolling value of the dominator. This is how $e \to g$ is assigned with 0 in the example illustrated in Figure 1.4. Further logic implications are then carried out based on the assigned values.

1.5 Mandatory Assignments and Recursive Learning

The fundamental component of ATPG is the logic implication procedure during fault excitation and propagation. Given a Boolean network and a set of preassigned values on some of the gates, logic implication is to derive the values of unassigned gates such that they are the same for any given input to the network that results in the preassigned values. Referring to the example in Figure 1.3, gate e is the first gate whose output is preassigned with a value (1). The unassigned gates are then determined by evaluating the truth tables of the gates with assigned values and the connectivity of the circuit.

The simplest form of logic implication is called direct implication, which is simply to evaluate the truth tables of the gates with assigned values to determine unknown values. However, it is not always possible to derive unknown values by truth table evaluation. For example, when the output of an AND gate is 0, which of its inputs should have the value 0? The problem of logic implication is actually much more complex than one would imagine.

The set of values on the unassigned gates found by the implication procedure is called the mandatory assignments (MAs) (or necessary assignments). A mandatory assignment is a

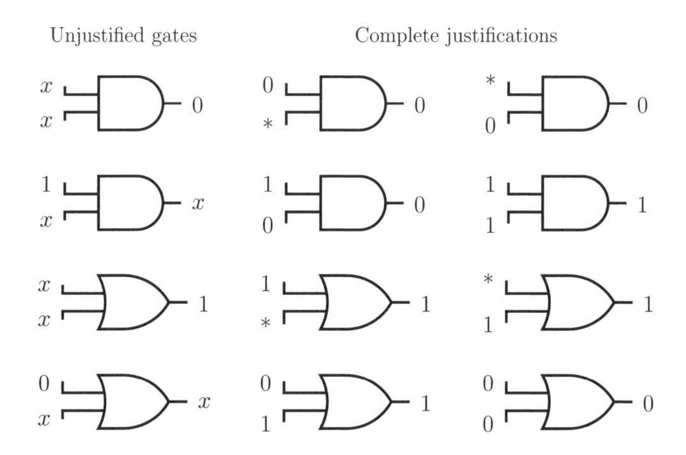

Figure 1.5 Unjustified AND/OR gates and their complete justifications

"forced mandatory assignment" (FMA) if it turns an originally testable fault to be untestable when its opposite value is used instead. With reference to Figure 1.3, the mandatory assignments for testing stuck-at-0 fault on wire $f \rightarrow g$ are $\{a = 0, b = 1, c = 1, d = 1, e = 0, f = 1\}$. Among these mandatory assignments, the mandatory assignment $b = 1$ is not an FMA because the fault $sa0(f \rightarrow g)$ is still testable even if $b = 0$ is used.

Recursive learning (Kunz and Pradhan 1994) or indirect implication is a method to derive all mandatory assignments for a given configuration of value assignments on the Boolean network. It is based on propagating the values found by direct implication according to the structural information of the circuit. The following gives several definitions that are needed in the recursive learning method:

1. Given a gate g that has at least one specified input or output signal, g is unjustified if there are one or more unspecified input or output signals of g for which it is possible to find a combination of value assignments that yield a conflict at g. Otherwise, g is justified. Figure 1.5 shows all possible assignments on a two-input AND gate and a two-input OR gate that will result in the gate becoming unjustified. In the figure, x represents unknown values and $*$ represents either 0 or 1.
2. A set of signal assignments $J = \{f_1 = V_1, f_2 = V_2, \ldots, f_n = V_n\}$, where $f_1, \ldots, f_n$ are the unspecified input or output signals of an unjustified gate g, is called the justifications for g if the combination of value assignments in J makes g justified.
3. Let C^* be the set of all justifications for gate g. The set of complete justification C for g is defined to be $C = \{J_i | J_i \subseteq J^*, \forall J^* \in C^*\}$.

An example of recursive learning in action is illustrated in Figure 1.6. Suppose the output of gate $g3$ is assigned with 1 (Figure 1.6(a)). After propagation of this value and justification of the assigned values, there can be three outcomes as shown in Figure 1.6(b)–(d). The common value assignment among the three configurations is $\{b = 1\}$. As a result, we can conclude that $g3 = 1$ implies $b = 1$ (Figure 1.6(e)). If only direct implication is applied and without recursive learning, this result cannot be obtained.

The level of logic implication is deeper through the use of recursive learning. As we shall see in later chapters, a deeper logic implication level is beneficial to rewiring techniques in that it allows them to identify more alternative wires. This has been proven by experimental results. The tradeoff of adopting recursive learning in rewiring algorithms is an increase in the runtime.

1.6 Graph Theory and Boolean Circuits

Definition 1.1 *Given a Boolean network, a cut-set (or cut for brevity) of a given node set O_v is a set of nodes not in O_v and are directly connected to the boundary nodes of O_v.*

It can be understood that any path passing through O_v to any PO must also go through some nodes in the cut-set. Basically, a cut of a given node v can be considered a "cutting boundary" of nodes that partitions the fanout cone of v from the node v. A cut-set is *K-feasible* if the cardinality of the cut-set is no more than K.

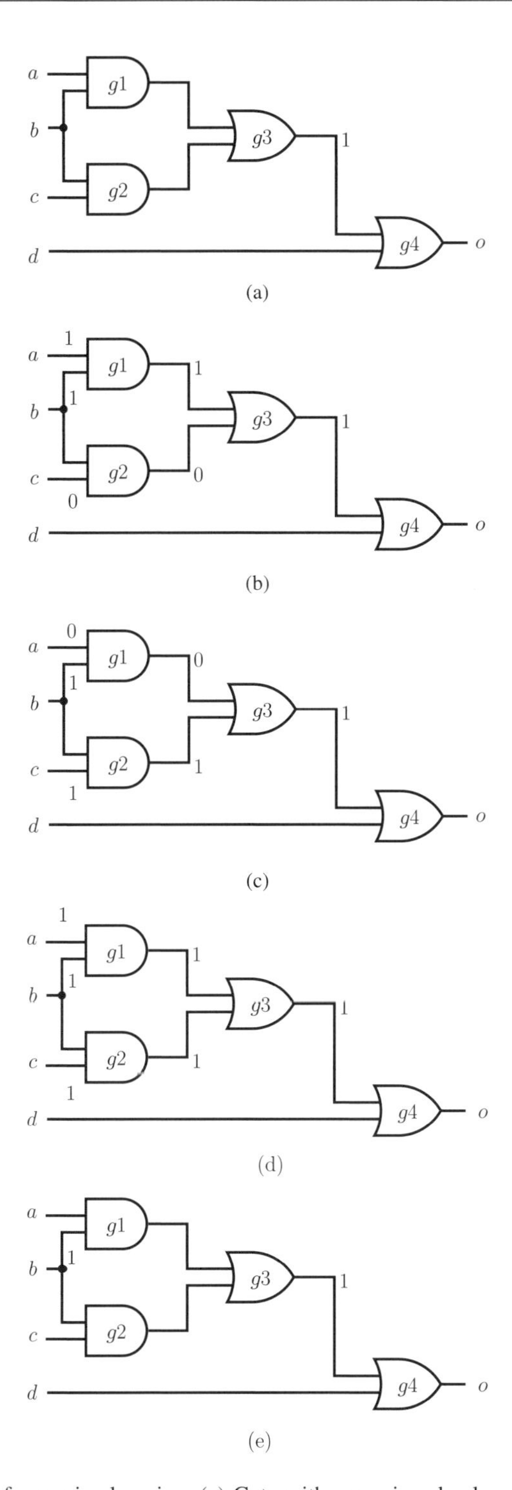

Figure 1.6 Example of recursive learning. (a) Gate with preassigned value: $g3$; (b) justifications (I) for $g1$, $g2$, and $g3$; (c) justifications (II) for $g1$, $g2$, and $g3$; (d) justifications (III) for $g1$, $g2$, and $g3$; (e) values implied from $g3 = 1$

Based on the above definitions, a *blocking cut* (B-cut) is defined as follows:

Definition 1.2 *Given a network, a cut-set S is a blocking cut (B-cut) of a node set if every path from the node set to the primary outputs must go through one and only one node in S.*

References

M. Damiani and G. De Micheli. Observability don't care sets and Boolean relations. In *Computer-Aided Design, 1990. ICCAD-90. Digest of Technical Papers, 1990 IEEE International Conference on*, pages 502–505, November 1990. doi: 10.1109/ICCAD.1990.129965.

N. K. Jha and S. Kundu. *Testing and Reliable Design of CMOS Circuits. Kluwer International Series in Engineering and Computer Science*. Kluwer Academic Publishers, 1990. URL http://books.google.com.hk/books?id=T0lTAAAAMAAJ.

T. Kirkland and M. R. Mercer. A topological search algorithm for ATPG. In *DAC '87: Proceedings of the 24th ACM/IEEE Design Automation Conference*, pages 502–508, New York, 1987. ACM. ISBN: 0-8186-0781-5. doi: 10.1145/37888.37963.

W. Kunz and D. K. Pradhan. Recursive learning: a new implication technique for efficient solutions to CAD problems-test, verification, and optimization. *IEEE Transactions Computer-Aided Design*, 13:1143–1158, 1994.

2

Concept of Logic Rewiring

2.1 What is Rewiring?

Rewiring is a technique to replace certain existing wires in a circuit with some other additional wires without affecting the circuit function. For example, consider the circuit in Figure 2.1(a).

Suppose the wire $a \rightarrow e$ is the wire that we want to remove because of delay violations or any other reasons. We can transform the circuit in the following way such that the original function of the circuit remains unchanged.

1. Connect f to the OR gate $a1$ (Figure 2.1(b)).
2. Disconnect the wire $e \rightarrow g$ and connect e to $a1$.
3. Remove the wire $a \rightarrow e$. Since the OR gate e has only one input now, it can also be removed.
4. Connect b to $a1$ directly.

The modified circuit is shown in Figure 2.1(c).

It is easy to prove that the circuits before and after rewiring are logically equivalent and both of them implement the same Boolean function.

Wire $a \rightarrow e$ is called a target wire; the wire that substitutes wire $a \rightarrow e$ is call an alternative wire. In this case, the alternative wire is $f \rightarrow a1$, or $f \rightarrow e(AND)$, where the latter means connecting f and e to an AND gate. There are cases in which an inverter has to be added in the alternative wire. We will use $a \nrightarrow b$ to represent a wire connecting a and b through an inverter.

Many researches have been working on rewiring techniques over the years. Rewiring techniques can be classified into three major classes: namely ATPG-based, SPFD-based, and graph-based rewiring. ATPG-based rewiring includes redundancy addition and removal for multilevel Boolean optimization (RAMBO), REWIRE, RAMFIRE. The recently developed and more general class of ATPG-based rewiring is the error-cancellation-based rewiring, including error-cancellation rewiring (ECR), flow-graph-based error cancellation rewiring (FECR), and cut-based error cancellation rewiring (CECR). SPFD-based local rewiring and the more general SPFD-based global rewiring form the SPFD-based family. GBAW is the representative of graph-based rewiring. Brief reviews on the various rewiring techniques are given in the following text.

Boolean Circuit Rewiring: Bridging Logical and Physical Designs, First Edition.
Tak-Kei Lam, Wai-Chung Tang, Xing Wei, Yi Diao and David Yu-Liang Wu.

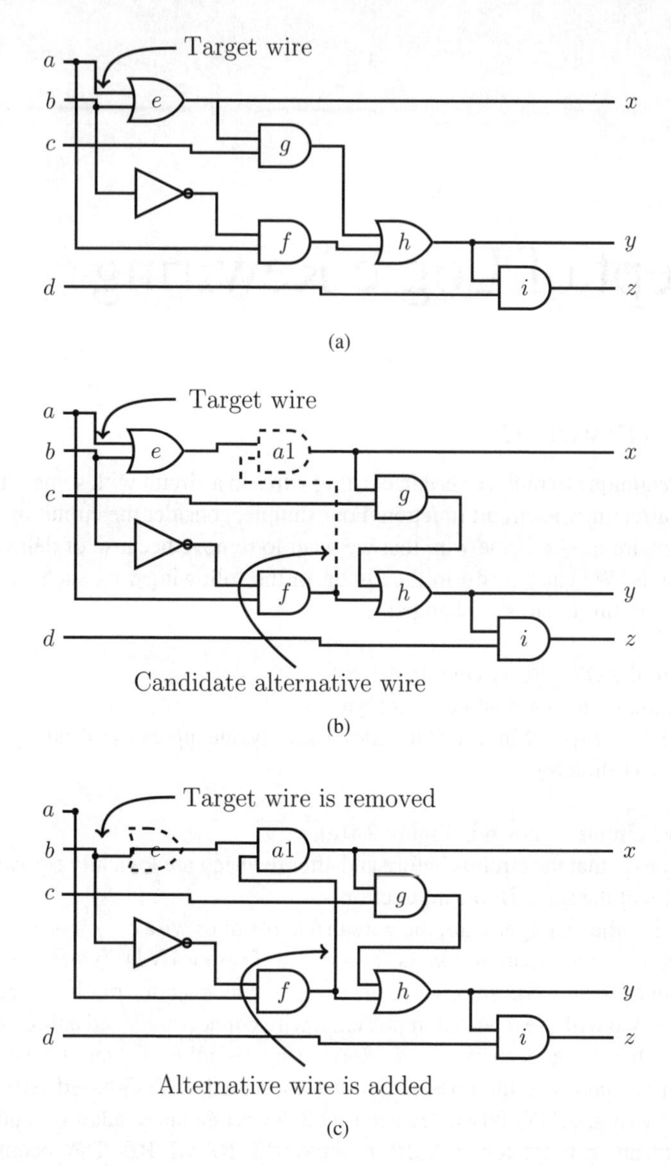

Figure 2.1 Example of rewiring. (a) Original circuit where a target wire is selected; (b) addition of an alternative wire; (c) rewired circuit

2.2 ATPG-based Rewiring Techniques

2.2.1 Add-First

2.2.1.1 RAMBO

Two decades ago, in Cheng and Entrena (1993) and (Entrena and Cheng 1995), the authors proposed the first rewiring technique called RAMBO. In RAMBO, the first step is to identify an alternative wire w_a whose addition to a circuit will make some originally irredundant target wires w_{t_i} in the circuit to become redundant. If w_a itself is found to be redundant, it will be added into the circuit. The redundant wires w_{t_i} can then be removed. In essence, this flow is the same as what we have just presented at the beginning of this chapter. The rewiring techniques following this methodology is known as the add-first rewiring techniques to reflect the fact that the addition of redundant alternative wires is the first step in the rewiring process. Because of the characteristics of the flow, they are also known as redundancy-addition-and-removal-based (RAR-based) rewiring techniques.

RAMBO was originally designed for area optimization (gate and wire count reduction). In the optimization flow proposed by the authors, as many redundant wires as possible are added into a given circuit first. The circuit is then evaluated to identify the redundant wires. The newly created redundant wires are then removed finally.

In the RAMBO approach, redundant wires to be added to a circuit are identified by logic implication. An example of using logic implication to identify new redundant wires is shown in Figure 2.2(a) (Cheng and Entrena 1993). The values of the primary inputs a, b, and c are all assigned with 1. Then, the output of gate $g2$ is 0 because its input $\bar{a}$ is 0. We can say that $a = 1$ implies $g2 = 0$. Since $g1 = 1$ implies $a = 1$, $g1 = 1$ implies $g2 = 0$. Based on this analysis, we can derive that connecting $g1$ to $g2$ does not change the function implemented by gate $g2$ and the new wire $g1 \rightarrow g2$ is redundant. Similarly, the new wire $g3 \rightarrow g1$ is redundant too.

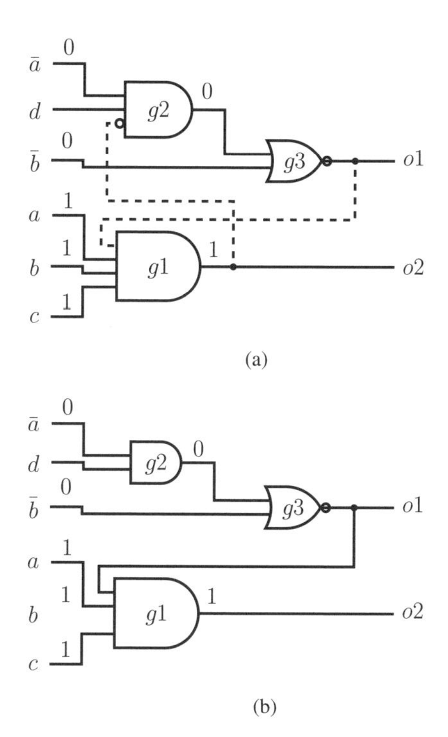

Figure 2.2 Redundancy addition and removal. (a) Identification of redundant wires; (b) addition of wire $g3 \rightarrow g1$ and removal of wire $b \rightarrow g1$

After the addition of redundant wires, the redundancy of every wire in the original circuit is checked using stuck-at fault tests (Section 1.2). It can be found that the addition of wire $g1 \rightarrow g2$ does not make any wires in the original circuit redundant. However, the addition of wire $g3 \rightarrow g1$ makes $b \rightarrow g1$ redundant. The resulting circuit after removing wire $b \rightarrow g1$ is shown in Figure 2.2(b).

2.2.1.2 REWIRE

Since its invention, RAMBO has been greatly improved by later researches. Various gradual improvements were published in Chang et al. (1996), Chang and Marek-Sadowska (2001a), Chang et al. (1999), Chang and Marek-Sadowska (1996), and Entrena and Cheng (1995). The major improvement of these techniques (REWIRE) over RAMBO is the redundancy checking mechanism. Theories that tell whether a wire is irredundant after the addition of alternative wires were derived. Hence, those wires that will not become redundant for sure can be skipped for evaluation. It can be expected that REWIRE runs much faster than RAMBO because redundancy checking consumes the most amount of time in rewiring.

As we can see from the simple example in Section 2.2.2.1, the earliest rewiring technique RAMBO performs redundancy checks that may be unnecessary. For instance, wire $d \rightarrow g2$ in Figure 2.2(a) can never be redundant after the addition of any one of the two new redundant wires. Its redundancy does not have to be evaluated at all. RAMBO may also add useless and redundant wires in the aspect of reducing the number of gates and wires. An example of this is the new redundant wire $g1 \rightarrow g2$, which does not cause any wires in the original circuit to be redundant.

The concepts of dominators (Section 1.4) and mandatory assignments (MAs) (Section 1.5) are heavily applied in REWIRE. Basically, researchers have theorized two major kinds of redundant alternative wires whose addition will make some wires redundant in the original circuit. For a target wire and its corresponding fault, one kind of its alternative wires can be identified by the dominators of its fault and another kind can be identified by the forced mandatory assignments (FMAs) (Section 1.5). We use the following example to introduce the ideas.

Consider, for example, the circuit shown in Figure 2.3(a) in which the dotted wire $c \rightarrow g2$ exists in the circuit and is the target wire. In order to identify the alternative wires of the target wire, a stuck-at fault test targeting $sa1(c \rightarrow g2)$ is first performed. The mandatory assignments are derived as indicated. Gate $g5$ is a dominator of the fault $sa1(c \rightarrow g2)$. Since its value is $0/1$, if we can force the faulty value to change from 1 to 0, the fault $sa1(c \rightarrow g2)$ will not be observable. The value of gate $g1$ is 0 because both of its inputs c and d are 0. Connecting $g1 \rightarrow g5$ can therefore achieve our goal. Furthermore, it can be easily proven that $g1 \rightarrow g5$ is redundant. Its addition to the circuit makes $c \rightarrow g2$ redundant and is therefore an alternative wire. From this example, we can understand how alternative wires are found using the concept of dominators.

FMA is another major concept that is useful for the determination of alternative wires. The target wire $g1 \rightarrow g5$ in Figure 2.3(b) is indicated by the dashed line. Similar to the previous example, a stuck-at fault test is first performed. Gate $g1$ has to be assigned with 0 to activate the fault $sa1(g1 \rightarrow g5)$. Then, the primary inputs c and d should be 0. Gate g5 is a dominator of $sa1(g1 \rightarrow g5)$, so its side inputs $g3 \rightarrow g5$ and $e \rightarrow g4$ have to be assigned with its noncontrolling value 1. Hence, both the primary input e and gate $g2$ should be 1. Finally, by backward implication again, we can reason that the primary inputs a and b must be assigned with 1. It was proven that all mandatory assignments obtained by backward implication are

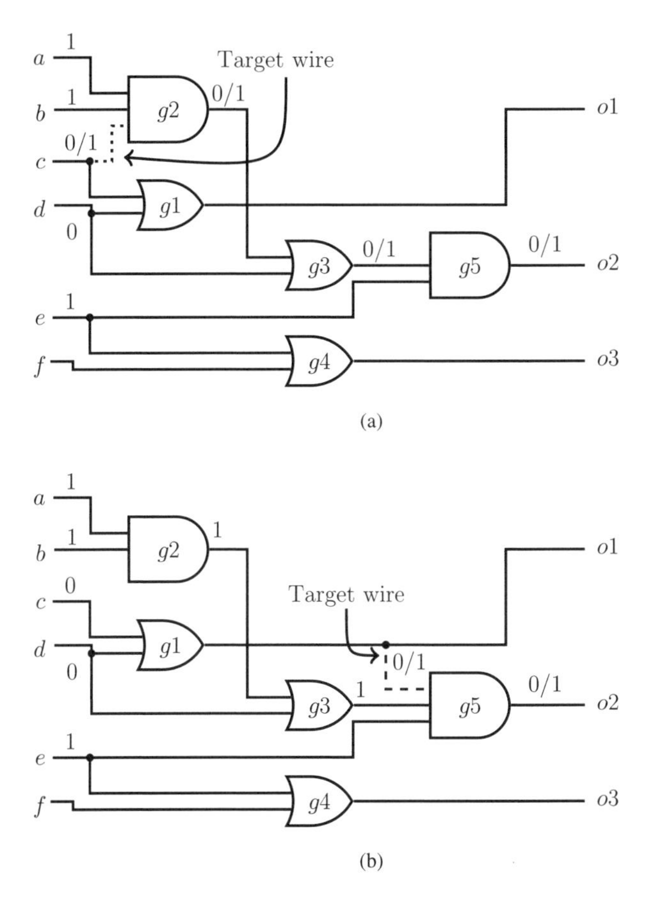

Figure 2.3 Identification of alternative wires in REWIRE. (a) Identification of alternative wires by dominators; (b) identification of alternative wires by FMAs

FMAs. Thus, $g2 = 1$ is a forced mandatory assignment. The definition of FMA of a fault is that it is a mandatory assignment whose value must be preserved so as to test the fault. We know it intuitively that forcing an FMA to have its opposite value can make the fault untestable. This analysis suggests connecting $c \to g2$, which will flip the value of gate $g2$. Wire $c \to g2$ is found to be redundant and is therefore an alternative wire for the target wire $g1 \to g5$.

Figure 2.3 summarizes the two major concepts, namely dominators and FMAs, that are useful for alternative wire identification in REWIRE. From this figure, readers may have realized that the pairs of target wire and alternative wire in Figure 2.3(a) and (b) are the same. Indeed, this is correct. It was proven that mutual redundant wires are the target and alternative wires of one another. Figure 2.4 shows another such example.

2.2.1.3 RAMFIRE

There have been other attempts aiming to reduce the number of redundancy checking. RAMFIRE (Chang and Marek-Sadowska 2001b, Change et al. 2003) is another ATPG-based

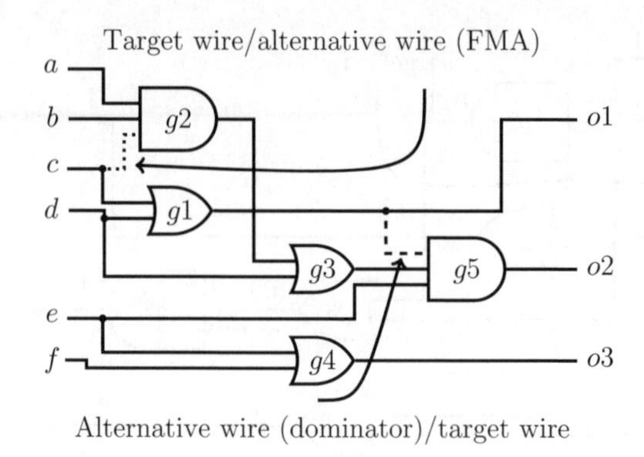

Figure 2.4 Mutual target and alternative wires

rewiring algorithm. It utilizes a polynomial-time redundancy identification technique called FIRE (Iyer and Abramovici 1996) to reduce unnecessary redundancy checking.

In FIRE, the concepts of uncontrollability and unobservability are applied. An uncontrollable status $s = \bar{0}$ ($s = \bar{1}$) of a signal s means that the signal is uncontrollable for $0(1)$; that is, that signal cannot assume the value of $0(1)$. The uncontrollability status of a signal can be propagated throughout the circuit by implication. Figure 2.5 shows the implication rules for uncontrollability. The arrows indicate the directions of implications. For example, if an AND gate's output cannot be set to 0, none of its inputs can be set to 0 either; that is, all input signals are assigned with 1.

The information about uncontrollability is useful for the determination of unobservability. If one of a gate's inputs cannot be set to the noncontrolling value of the gate, any stuck-at faults on the other inputs of the gate cannot be observed. For example, if the unobservability of one of the inputs of a two-input AND gate is $\bar{1}$, the AND gate always outputs 0, and both stuck-at-0 and stuck-at-1 on other inputs are then unobservable. As a matter of fact, if a wire is unobservable, both the stuck-at-0 and stuck-at-1 faults on that wire are not testable.

The implication rules for unobservability for AND, OR, and NOT gates are depicted in Figure 2.6, where the symbol $*$ denotes that the signal is unobservable. As in Figure 2.5, the

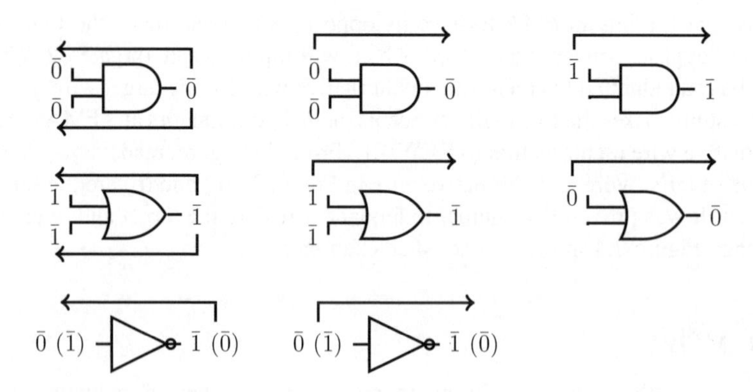

Figure 2.5 Uncontrollability implication for AND, OR, and NOT gates

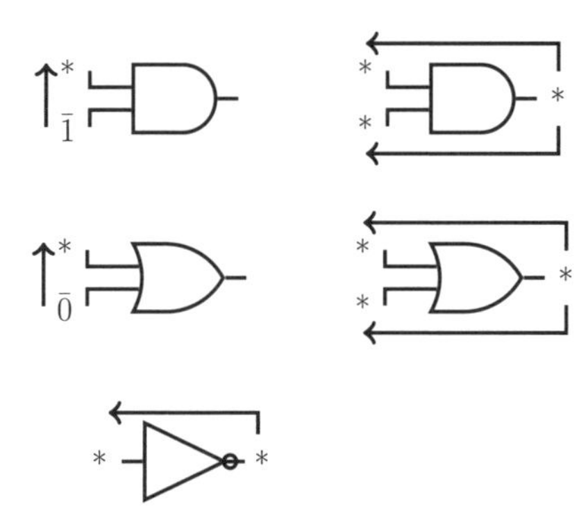

Figure 2.6 Unobservability implication for AND, OR, and NOT gates

directions of implications are indicated by the arrows. Since AND, OR, and NOT gates can implement any Boolean functions, the rules for other types of gates can be easily derived from those for AND, OR, and NOT gates. From the rules, it can also be seen that unobservability always propagates backwards.

We can make use of the information about uncontrollability and unobservability to identify redundant faults and redundant wires. The concepts are as follows:

(i) Since it is required that a wire can be assigned with value 1 (0) so as to test the stuck-at-0 (stuck-at-1) fault on the wire (to achieve fault excitation), if the uncontrollability of the wire is implied to be $\bar{1}$ ($\bar{0}$), the stuck-at-0 (stuck-at-1) fault on the wire cannot be tested and is redundant.

(ii) If a wire is unobservable, both stuck-at-0 and stuck-at-1 faults on the wire are not testable as well as the wire itself.

The method to identify redundancy by utilizing the information about uncontrollability and unobservability starts by assigning the uncontrollability of a wire w to be $\bar{0}$ ($\bar{1}$). After propagating the value by implication, two sets of uncontrollability and unobservability values implied on some other wires can be obtained. The set of value assignments resulting from the implication of $w = \bar{0}$ is denoted as $S(w = \bar{0})$ and that from the implication of $(w = \bar{1})$ is denoted as $S(w = \bar{1})$. Then, from $S(w = \bar{0})$ and $S(w = \bar{1})$, two sets of untestable faults $F(w = \bar{0})$ and $F(w = \bar{1})$ are, respectively, determined. The set of faults $F(w = \bar{0})$ ($F(w = \bar{1})$) consists of the faults whose necessary condition to be detectable is $w = \bar{0}$ ($w = \bar{1}$). The intersection of $F(w = \bar{0})$ and $F(w = \bar{1})$ is a set of faults that cannot be tested unless w is 0 and also 1 simultaneously. Clearly, this is impossible. Hence, the faults in the intersection of $F(w = \bar{0})$ and $F(w = \bar{1})$ are untestable and redundant under any conditions.

We mentioned that the ATPG-based rewiring algorithm RAMFIRE utilizes the FIRE redundancy identification technique that has just been introduced. This rewiring algorithm performs rewiring by first adding an alternative wire w_a to the given circuit, and then performing a FIRE redundancy test on w_a. The set of untestable faults $F(w_a = \bar{0})$ and $F(w_a = \bar{1})$ are obtained. Finally, the target wires w_t that have become redundant after addition of the alternative wire can be identified.

Compared to RAMBO, RAMFIRE needs no redundancy test for every wire. Therefore, besides REWIRE, RAMFIRE is also substantially faster than RAMBO. It was experimentally proven that REWIRE can find more alternative wires than RAMFIRE. In other words, REWIRE has a higher rewiring ability.

Definition 2.1 *Rewiring ability is defined as the ratio of the wires in a circuit that has alternative wires to the total number of wires.*

2.2.2 Delete-First

In the previous section, we briefly introduce the add-first rewiring techniques. Is it necessary that we have to add redundant wires to force some originally irredundant wires to be redundant? Latest research tells us that the requirement is not a must. In fact, irredundant wires can also be added to a circuit to turn some wires in the circuit into redundant wires. Furthermore, instead of adding alternative wires to remove a target wire, a target wire can be removed first to determine the alternative wires required to maintain the original circuit function. We call this methodology the delete-first rewiring, or error-cancellation-based rewiring.

Delete-first rewiring can be best explained by the simple examples shown in Figure 2.7. For the AND gate, the uppermost wire (target wire) is to be removed. Since its value is 0, its removal causes an error as if it is replaced by a constant 1 (the noncontrolling value of AND gates). This can be modeled as a stuck-at-1 fault. On the other hand, the bottom most wire (alternative wire) is a newly added wire whose value is 0. The AND gate can be assumed to have an input with value 1 originally. Then, the value of the input changes from 1 to 0 after the

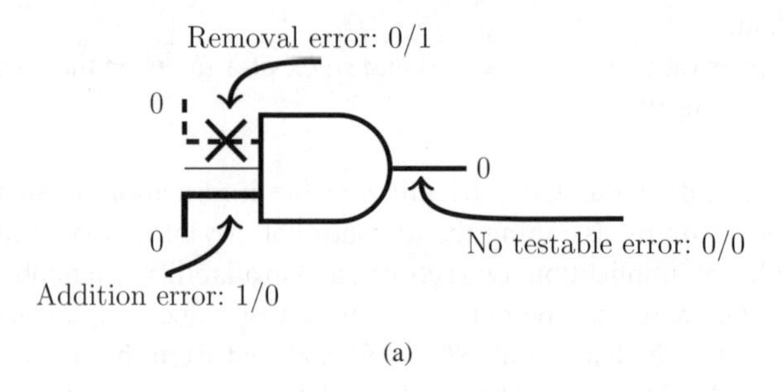

(a)

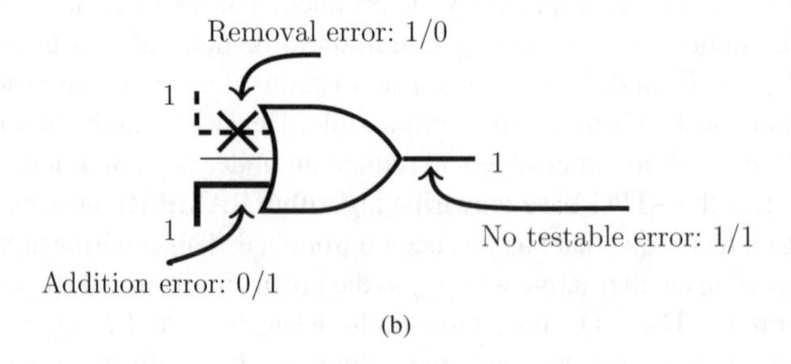

(b)

Figure 2.7 Examples of delete-first rewiring/error cancellation

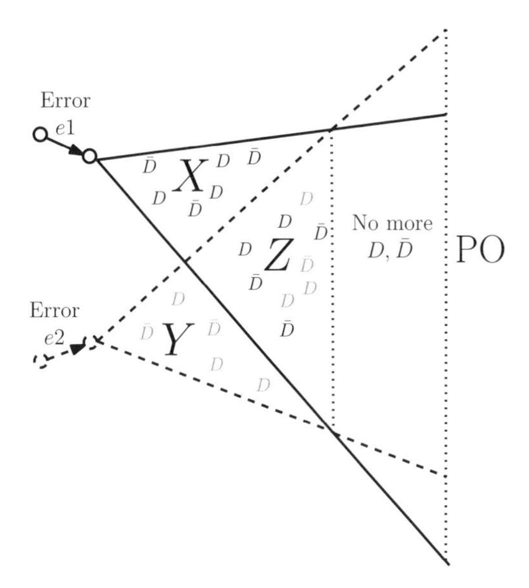

Figure 2.8 General abstract view of error cancellation

addition of the alternative. The error induced can be modeled as a stuck-at-0 fault. Although there are errors caused by the removal of the target wire and the addition of the alternative wire, the errors cancel each other eventually as indicated in the figure. The original circuit function is therefore maintained after rewiring. Regarding OR gate and other types of gates, the analysis is similar.

A more general abstract view of error cancellation is depicted in Figure 2.8. In the figure, $e1$ and $e2$ are two errors in a circuit. The effect of error $e1$ can be observed in zone X only, whereas the effect of error $e2$ can be observed in zone Y only. The errors meet in zone Z where their effects cancel each other. As a result, there exist errors but they are unobservable at the outputs of the circuit.

The idea of delete-first rewiring or error cancellation can be traced back to (Muroga et al. 1989). It was first discussed theoretically in detail in Chang and Marek-Sadowska (2007). The authors derived theories that generalize the add-first rewiring methodology into the delete-first rewiring methodology. However, direct implementation of the theories and algorithms are computationally intensive because they are general. More practical but less general delete-first rewiring techniques are therefore designed.

2.2.2.1 IRRA

If we can afford to remove a target wire at any cost, it is unnecessary for us to use an alternative wire to substitute the target wire. As an extreme case, the original circuit shown in Figure 2.9(a) can be replaced by the logic implemented by gates j and k, which are the duplicates of gate f and the NOT gate connecting with it (Figure 2.9(b)). Clearly, adopting this approach can increase the rewiring ability of a rewiring technique. The logic to be added, however, may be too costly and undesirable.

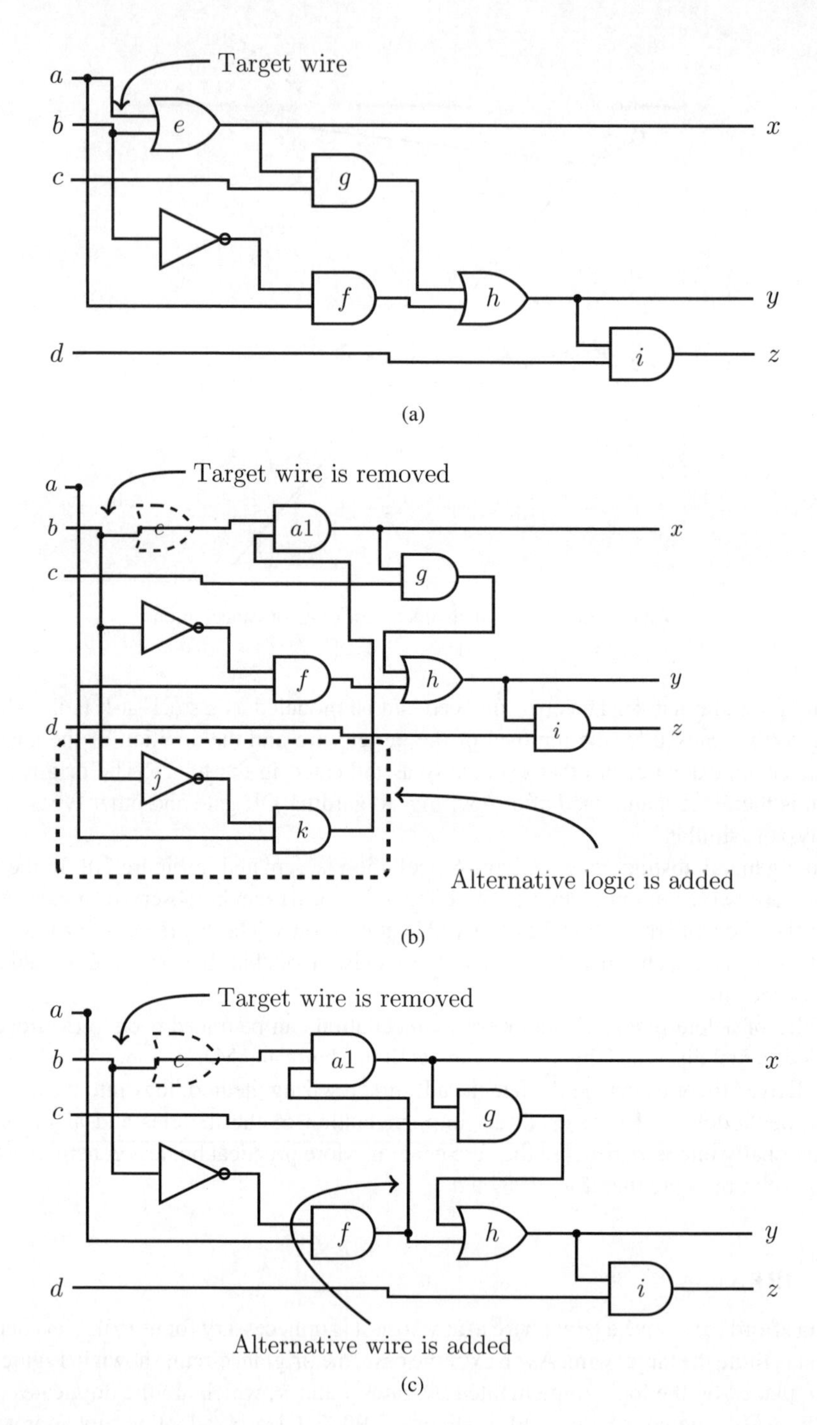

Figure 2.9 Example of rewiring by logic addition. (a) Original circuit; (b) alternative circuit; (c) simplified alternative circuit

The authors in Lin and Wang (2009) suggested an innovative approach, called irredundancy removal and addition (IRRA), to perform rewiring by adding an alternative logic that can be reduced to a wire. In other words, the alternative logic is allowed to be added temporarily into the circuit as the source node of an alternative wire, provided it can be simplified in the end, even if it is very complex. In the our current example, the temporary alternative logic can obviously be reduced to a wire, as shown in Figure 2.9(c).

2.2.2.2 ECR

Having adopted the idea of temporary alternative logic, the choice of alternative wires in IRRA is still limited. The reason is that, for a given target wire, IRRA tries to use only the dominators and nodes with FMAs as the destination nodes of the alternative wires. This practice is the same as what the add-first rewiring techniques normally do. If the destination node of an alternative wire is a dominator, the alternative wire is redundant; if it is a node with FMA, it may be irredundant. We have to expand the solution space for identifying irredundant wires.

Considering this limitation, the authors of ECR (Yang et al. 2010) and (Lam et al. 2012) proposed a rewiring technique that advances one more step toward the theoretical limit. In their approach, instead of using only the dominators and nodes with FMA, the dynamic dominators of a fault are also considered to be the destination nodes of the alternative wires.

Dynamic dominators

Different from the dominators that have been introduced in Section 1.4 , the dynamic dominators (Krieger et al. 1991) of a fault are not only related to the structure of the circuit and cannot be determined structurally. A dynamic dominator of a stuck-at fault is defined to be a node in the circuit that the propagation of the fault must pass through so as to reach the primary outputs. It is obvious that the dynamic dominator has value $1/0 \quad (D)$ or $0/1 \quad (\bar{D})$ for the stuck-at fault test.

Figure 2.10 shows the dominators of a stuck-at fault $sa1(b \to g1))$, which is activated and propagated using the MAs as indicated. It can be seen that gates $g1$, $g3$, and $g4$ are the dominators of the fault because all paths containing $b \to g1$ have to pass through the gates. The path of fault propagation has to pass through $b \to g1 \to g3 \to g4$ and end at the PO o. Hence, the set of dominators are also dynamic dominators in this example.

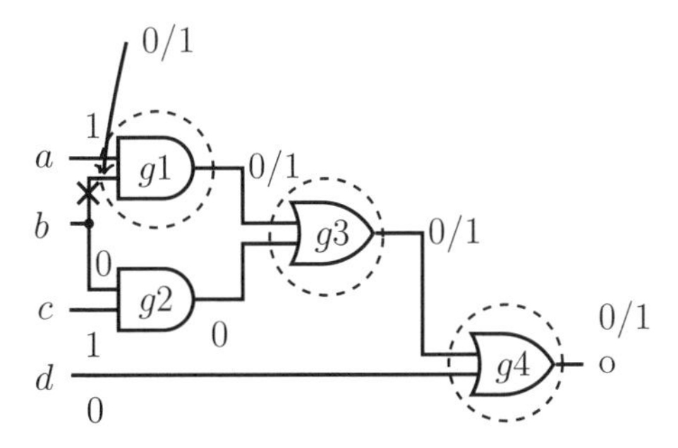

Figure 2.10 Examples of dynamic dominators

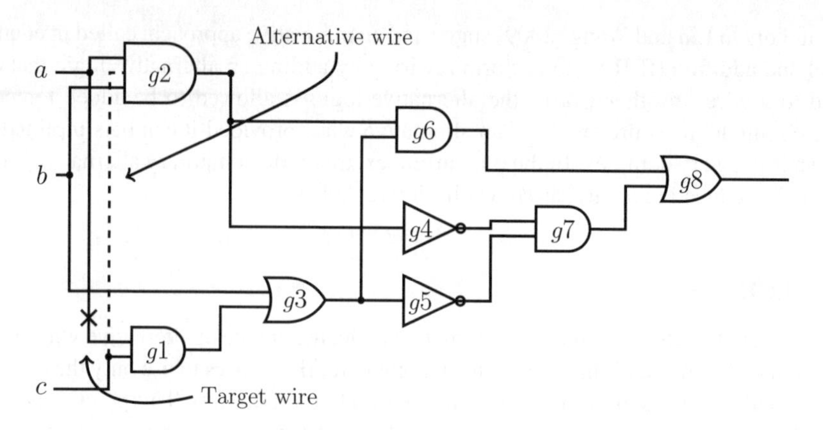

Figure 2.11 Example of an irredundant alternative wire

The use of dynamic dominators allows more irredundant alternative wires to be identified. The example shown in Figure 2.11 (Chang and Marek-Sadowska 2007) is a case where the traditional add-first rewiring techniques and IRRA cannot find the indicated alternative wire for the target wire. The reason is that the destination node $g2$ of the alternative wire is not a dominator of $sa1(a \rightarrow g1)$ and not a node having forced mandatory assignment either.

ECR also employs the strategy of using temporary alternative logic as the source nodes of alternative wires. However, ECR applies a more powerful technique (called node merging) to try to reduce temporary alternative logic to a wire. It has been proven empirically that ECR can obtain twice the number of alternative wires found by IRRA. The adoption of dynamic dominators as the destination nodes of alternative wires and a more effective method to simplify the alternative logic are the main contributions to the improvement.

2.2.2.3 FECR

Although the use of dynamic dominators can bring significant improvement in the identification of alternative wires, the maximum potential of the theory of error cancellation still cannot be utilized. The major practical rewiring techniques introduced thus far focus on using one alternative wire to replace one target wire. Since there are ways that make errors cancel each other, we do not actually need to limit ourselves to the scheme of "one-by-one rewiring."

For example, in Figure 2.12, the stuck-at fault $sa0(b \rightarrow g1)$ is propagated through gate $g1$, then gate $g5$, and subsequently through gates $g7$ and $g8$ simultaneously. The sets of gates $\{g1\}$, $\{g5\}$, and $\{g7, g8\}$ are called errors frontiers (indicated as dashed snaky lines) with respect to the fault $sa0(b \rightarrow g1)$. Obviously, if the effect of the error can be cancelled at either $\{g1\}$, $\{g5\}$, or $\{g7, g8\}$, the fault is unobservable.

In the recently published rewiring technique FECR (Yang et al. 2012), the concept of error frontiers is adopted and the nodes in the error frontiers are considered during the process of identifying the destination nodes of alternative wires. In short, the error frontiers of some faults are a set of nodes through which the fault propagation can be blocked. They are determined by first modeling the circuit as a graph and calculating the minimum cuts of the fault propagation

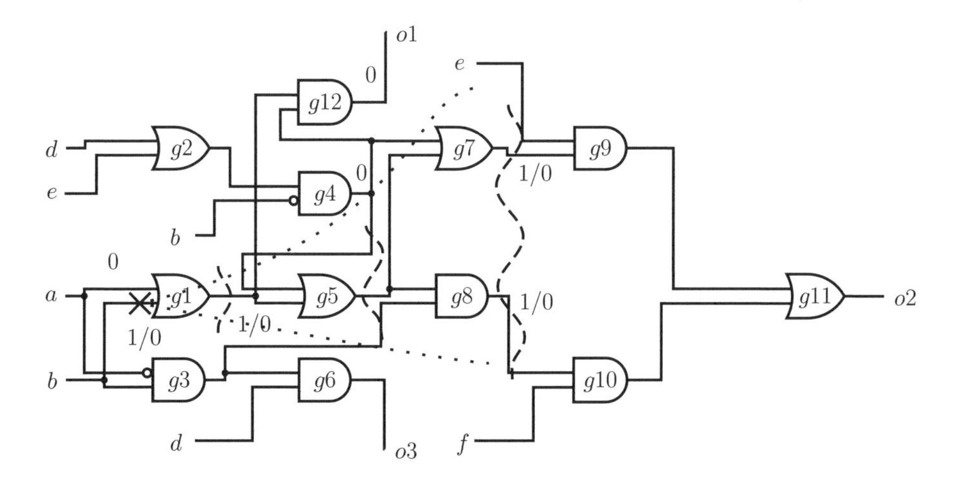

Figure 2.12 Error propagation of $sa0(b \rightarrow g1)$ and the error frontiers

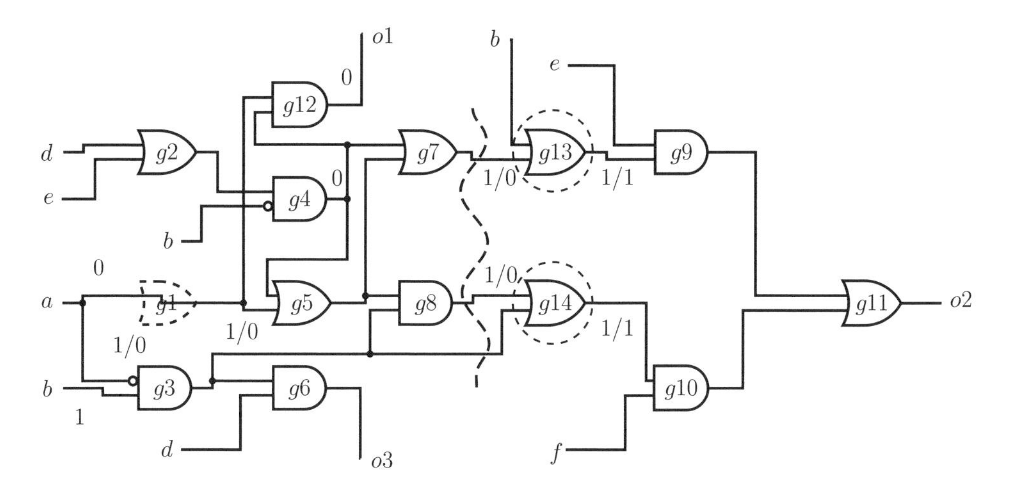

Figure 2.13 Multiple alternatives wires

paths in the graph (will be discussed in detail in later chapters). This idea is closer to the theory of error cancellation, and allows the theory to be implemented more practically. It is possible to substitute a target wire with multiple alternative wires then. As an example, the target wire $b \rightarrow g1$ in Figure 2.12 can be replaced by the two alternative wires $b \rightarrow g7(OR)$ and $g3 \rightarrow g8(OR)$ ($b \rightarrow g13$ and $g3 \rightarrow g14$), as shown in Figure 2.13, where the newly added OR gates are indicated by dashed circles.

The rewiring technique FECR also considers other nodes with FMAs as the destination nodes of alternative wires. For the source nodes, it follows the methodology of IRRA and ECR, which tries to add alternative logic first and reduce the alternative logic to a wire.

2.2.2.4 CECR

The major problem of FECR is its requirement of huge computation power due to the computation of minimum cuts. It is improved in (Wei et al. 2013), in which the authors suggested the CECR rewiring technique. In CECR, the fault propagation algorithm is extended to find more alternative wires. It achieves this by introducing more mandatory assignments so that there are more potential sources and destinations for alternative wires. An effective cut enumeration algorithm is used in CECR to reduce the complexity significantly. The authors also suggested a windowing approach to make their rewiring technique scalable and practical for large circuits. The rewiring ability of CECR is found to be at least as strong as that of FECR, but CECR requires much less computational resource.

2.3 Non-ATPG-based Rewiring Techniques

2.3.1 Graph-based Alternate Wiring (GBAW)

Rewiring techniques are mostly evolved around the notion of automatic test pattern generation. ATPG is fast and scalable, but its runtime may still be an issue especially when recursive learning is applied on complex circuits. In view of this, there have been attempts that tried to get rid of logic implication. The most innovative and representative class is graph-based rewiring. In Wu et al. (2000a), this idea was first proposed and known as GBAW. Later published works including Wu et al. (2000b), Chim and Wu (2007), Chim et al. (2009), and Chim et al. (2011) are derived from GBAW to improve its rewiring ability.

Graph-based rewiring techniques regard the problem of identifying alternative wires for a target wire as a problem of pattern matching in a graph. It models an input circuit as a directed acyclic graph (DAG) and tries to look for special patterns inside the graph. The first step in GBAW is to convert the given circuit to a DAG. The basic concepts of Boolean networks and DAGs and the notations have been explained in Section 1.1.

A special target pattern to look for is in fact a subcircuit structure that contains a target wire and the alternative wire for the target wire determined by other rewiring techniques. It is essentially a hard-coded configuration that tells the information of a sub-circuit structure: whether there is a pair of target and alternative wire, and how the alternative wire is connected. Three sample GBAW patterns are illustrated in Figure 2.14. Regarding each of the patterns, one can easily verify that the original circuit function, that after the removal of the target wire, and the addition of the alternative wire are the same.

The DAG representations of the three sample patterns are shown in Figure 2.14. In the figure, an edge may contains circle at the middle. This corresponds to the circuit schematic, and means that the value feeding into the destination node should be inverted. Above each node is a label consisting of its information: its gate type, the in-degree, and the out-degree. For instance, the node $g1$ in Figure 2.15(a) represents an AND gate whose number of fanins and fanouts are, respectively, 2 and 1. A node has a type of "DC" if the gate type is don't care (do not have to be taken into account). The same also applies to the in-degree and out-degree.

Then, the GBAW rewiring technique is to locate the preconfigured graph patterns in the DAG of the whole circuit. Since no logic implication is involved, GBAW and its variants can locate alternative wires with high speed. However, they are dependent on the richness and variety of the pattern library.

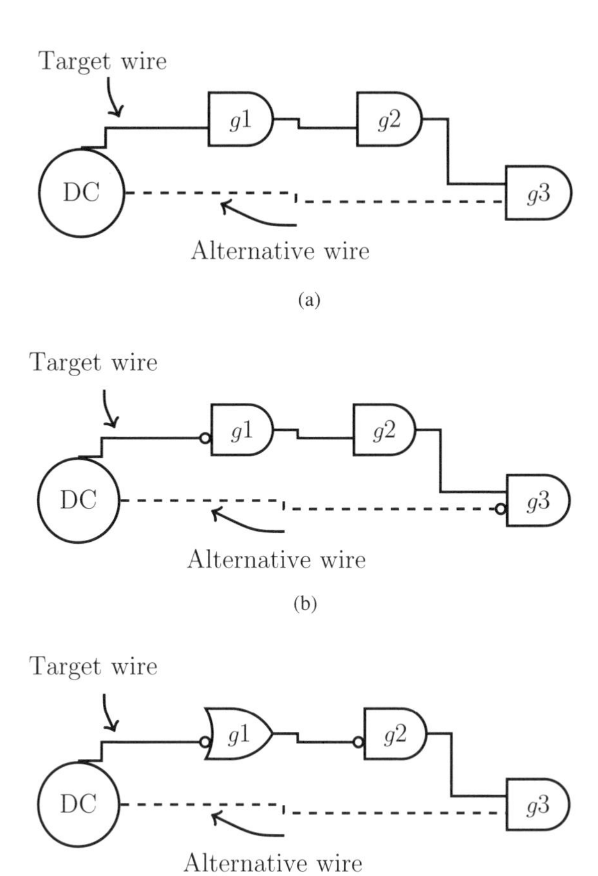

Figure 2.14 Sample graph-based alternate wiring patterns. (a) Sample pattern I; (b) II; (c) III

2.3.2 SPFD

SPFD-based rewiring technique is another class of rewiring techniques. This idea was first proposed in Yamashita et al. (1996). Since the idea of SPFD is rather complex, we present some essential terminologies before going into the methodology of alternative wire identification. The details and advanced material, as usual, will be left for later chapters.

A minterm of a function of n variables is the logical conjunction of all the variables in which each of the variable may be negated. Since there are 2^n number of combinations for n variables when negation is considered, a Boolean function of n variables has 2^n minterms. The truth table and all the minterms of a two-input AND function is listed in Table 2.1. If a minterm of a function evaluates to 1, it is in the on-set (ON_f) of the function. Otherwise, if it evaluates to 0, it is in the off-set (OFF_f) of the function. Thus, the set of minterms ab belongs to the on-set and the set of minterms $\{\bar{a}\bar{b}, \bar{a}b, a\bar{b}\}$ belongs the off-set of the two-input AND function. If a minterm of a function f does not have any effect on the function, it is a don't care and in the don't-care set (D_f). Therefore, for any given Boolean function, its minterms can be classified into three sets. A minterm can itself be viewed as a function.

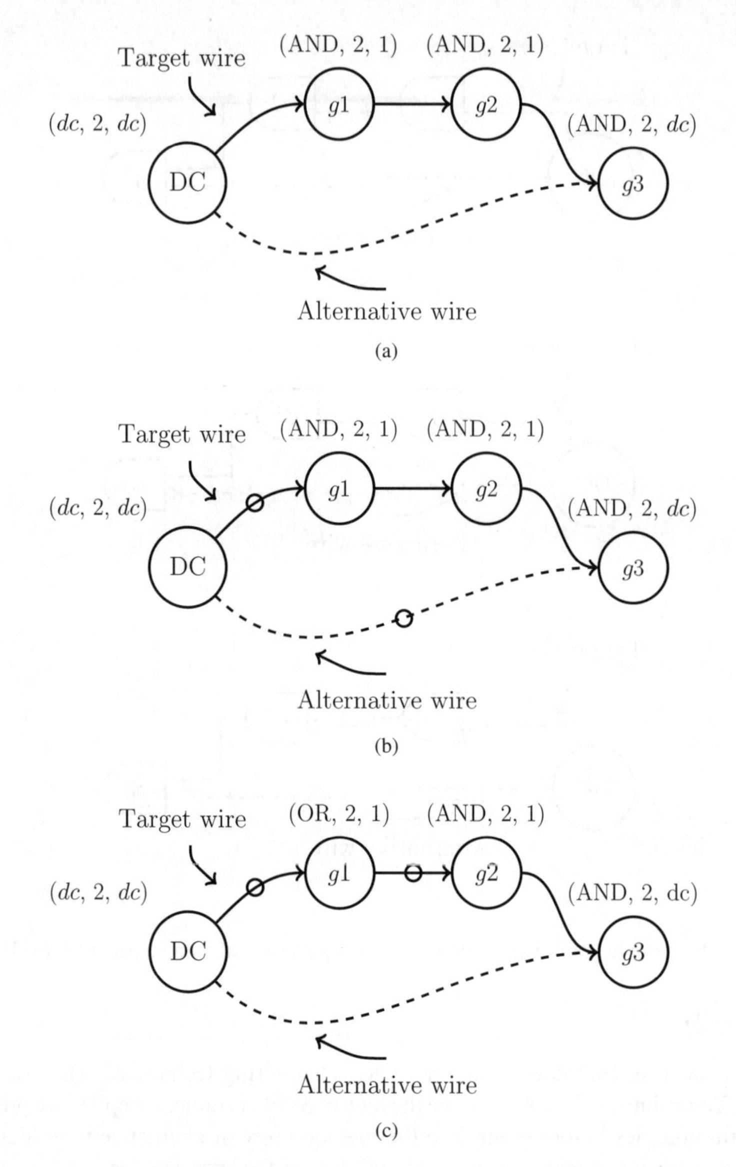

Figure 2.15 DAG representations of the sample GBAW patterns. (a) Sample graph pattern I; (b) II; (c) III

Two functions f and h are said to be distinguishable if:

(i) their on-sets are not empty,
(ii) their on-sets do not overlap with each other, and
(iii) there exists a function g that is 1 when f is 1 and is 0 when h is 1; or, is 1 when h is 1 and is 0 when f is 1.

Table 2.1 Truth table and minterms of logical AND function

a	b	$f(a,b) = a \cdot b$	Minterm
0	0	0	$\bar{a}\bar{b}$
0	1	0	$\bar{a}b$
1	0	0	$a\bar{b}$
1	1	1	ab

The first and second requirement are to ensure that there is no overlap between the on-sets of the two functions. The third statement means that the on-set of f has to be a subset of the on-set of g, and the on-set of g in turn has to be a subset of the on-set of $\bar{h}$. The function g is said to distinguish f and h.

The authors of Yamashita et al. (1996) cleverly interpret Boolean functions and express their flexibility using the concepts of minterms, on-sets and off-sets. Their idea is that any Boolean function can be implemented in any way as long as its minterms in the on-set and off-set are distinguishable from one another.

For a primary output implementing Boolean function f of a circuit, the Cartesian product of its on-set and offset forms a set of pairs of functions to be distinguished ($SPFD_f$). We use the notation $SPFD_f^{on}$ ($SPFD_f^{off}$) to represent the minterms in the on-set (off-set) of f. Consider the example circuit shown in Figure 2.16. The on-set and off-set of the functions implemented at the primary output gate g are listed in the first two columns of Table 2.2. The Cartesian product of this two sets of minterms consists of 63 pairs of functions, such as $ab\bar{c}\bar{d}$ and $\bar{a}b\bar{c}\bar{d}$, to be distinguished. The Karnaugh maps of $SPFD_g$, $SPFD_e$, $SPFD_f$, and $SPFD_i$ are shown in Figures 2.17–2.20.

For a node that is not a primary output in a circuit, its SPFD is derived from the SPFD of its output recursively. The SPFD of the output of a gate is the union of the SPFDs of its fanouts.

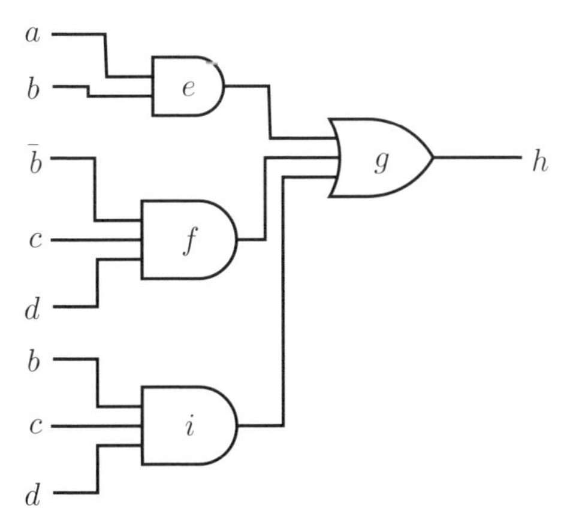

Figure 2.16 Implementation I of $h = ab + cd$

Table 2.2 SPFDs of the gates in Figure 2.16

$SPFD_g^{on}$	$SPFD_g^{off}$	$SPFD_e^{on}$	$SPFD_e^{off}$	$SPFD_f^{on}$	$SPFD_f^{off}$	$SPFD_i^{on}$	$SPFD_i^{off}$
$ab\bar{c}\bar{d}$	$\bar{a}b\bar{c}\bar{d}$	$ab\bar{c}d$	$\bar{a}b\bar{c}\bar{d}$	$\bar{a}\bar{b}cd$	$\bar{a}b\bar{c}\bar{d}$	$\bar{a}bcd$	$\bar{a}b\bar{c}\bar{d}$
$ab\bar{c}d$	$\bar{a}\bar{b}\bar{c}\bar{d}$	$ab\bar{c}d$	$\bar{a}\bar{b}\bar{c}\bar{d}$	$a\bar{b}cd$	$\bar{a}\bar{b}\bar{c}\bar{d}$	$abcd$	$\bar{a}\bar{b}\bar{c}\bar{d}$
$abcd$	$a\bar{b}\bar{c}\bar{d}$	$abcd$	$a\bar{b}\bar{c}\bar{d}$		$a\bar{b}\bar{c}\bar{d}$		$a\bar{b}\bar{c}\bar{d}$
$abc\bar{d}$	$\bar{a}b\bar{c}d$	$abc\bar{d}$	$\bar{a}b\bar{c}d$		$\bar{a}b\bar{c}d$		$\bar{a}b\bar{c}d$
$\bar{a}bcd$	$\bar{a}\bar{b}\bar{c}d$		$\bar{a}\bar{b}\bar{c}d$		$\bar{a}\bar{b}\bar{c}d$		$\bar{a}\bar{b}\bar{c}d$
$a\bar{b}cd$	$a\bar{b}\bar{c}d$		$a\bar{b}\bar{c}d$		$a\bar{b}\bar{c}d$		$a\bar{b}\bar{c}d$
$\bar{a}\bar{b}cd$	$\bar{a}bc\bar{d}$		$\bar{a}bc\bar{d}$		$\bar{a}bc\bar{d}$		$\bar{a}bc\bar{d}$
	$a\bar{b}c\bar{d}$		$a\bar{b}c\bar{d}$		$a\bar{b}c\bar{d}$		$a\bar{b}c\bar{d}$
	$\bar{a}\bar{b}c\bar{d}$		$\bar{a}\bar{b}c\bar{d}$		$\bar{a}\bar{b}c\bar{d}$		$\bar{a}\bar{b}c\bar{d}$

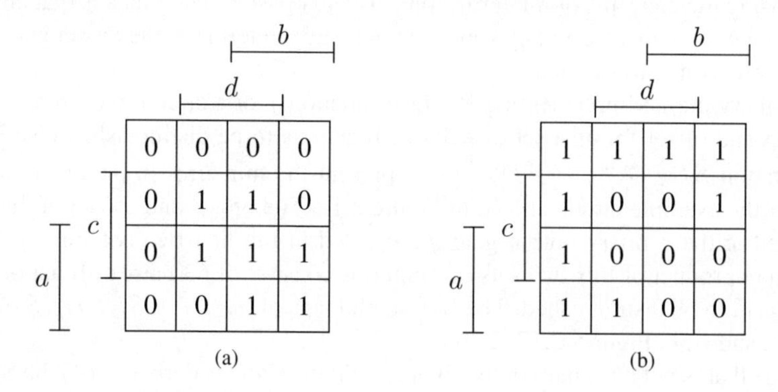

Figure 2.17 Karnaugh maps of $SPFD_g$. (a) $SPFD_g^{on}$; (b) $SPFD_g^{off}$

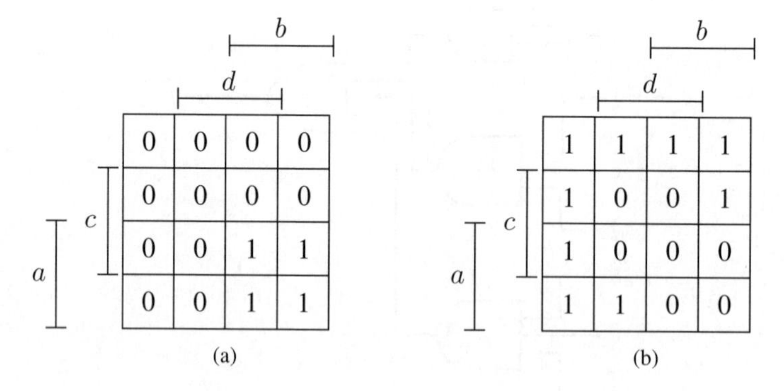

Figure 2.18 Karnaugh maps of $SPFD_e$. (a) $SPFD_e^{on}$; (b) $SPFD_e^{off}$

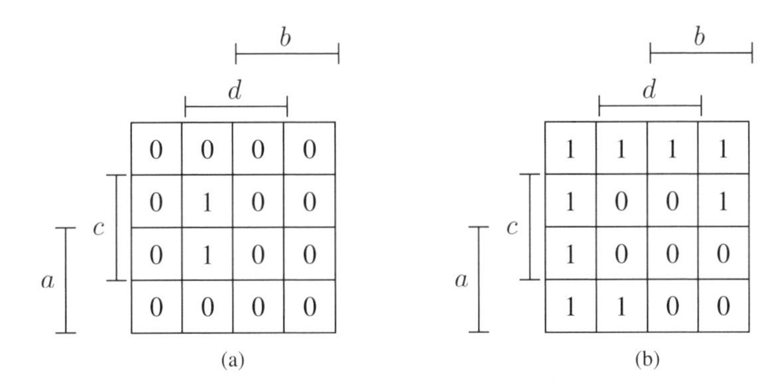

Figure 2.19 Karnaugh maps of $SPFD_f$. (a) $SPFD_f^{on}$; (b) $SPFD_f^{off}$

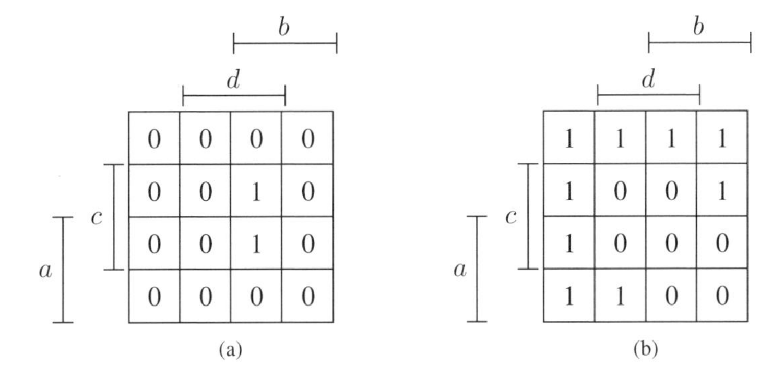

Figure 2.20 Karnaugh maps of $SPFD_i$. (a) $SPFD_i^{on}$; (b) $SPFD_i^{off}$

The SPFDs of the inputs of a gate are obtained by distributing the SPFD of the gate's output. For example, $SPFD_e^{on}$ contains the first four minterms of $SPFD_g^{on}$, and $SPFD_f^{on}$ and $SPFD_i^{on}$ contain the remaining minterms (listed in the third to eighth columns in Table 2.2). For simplicity, all $SPFD^{off}$ are the same in this example. This is not necessary. The rule of SPFD distribution is that the union of the SPFDs of the inputs must be equal to the SPFD of the output.

It can be easily proved that the functions ab, $\bar{b}cd$, and bcd can, Respectively, distinguish $SPFD_e$, $SPFD_f$, and $SPFD_i$. Therefore, $h = ab + cd$ can be implemented as shown in Figure 2.16. Since the distribution of SPFD from a gate's output to its inputs is flexible, we can use any function that can distinguish the SPFDs. Obviously, the simpler and direct implementation illustrated in Figure 2.21 is another option. Table 2.3 lists the SPFDs of the gates. The new SPFDs are shown in Figures 2.22 and 2.23.

Having understood the idea of SPFD and the mechanism of using it to manipulate function flexibility, the mechanism of alternative wire identification can be explained as a redistribution of SPFDs. The circuit shown in Figure 2.16 can be transformed into that shown in Figure 2.21 by removing wires $f \rightarrow g$ and $i \rightarrow g$, and then adding wire $i \rightarrow g$, as depicted in Figure 2.24. The removal of the target wires results in the inequality between the SPFDs of the inputs and the SPFD of the output of gate g; some pairs of functions in $SPFD_g$ cannot be distinguished. The function implemented by gate j is found to be able to distinguish the unsatisfied pairs of functions. It is thus added as an input of gate g so that the circuit function is preserved.

SPFD-based rewiring was also suggested in Yamashita et al. (1996). It has been improved over the years and several publications are available: Sinha and Brayton (1998), Cong et al. (2002a), Cong et al. (2002b) and Maidee and Bazargan (2007). The improvements include the ability to use multiple alternatives wires to replace a target wire ("one-by-many rewiring"). In theory, SPFD-based rewiring can be extended to ";many-by-many rewiring" natively just

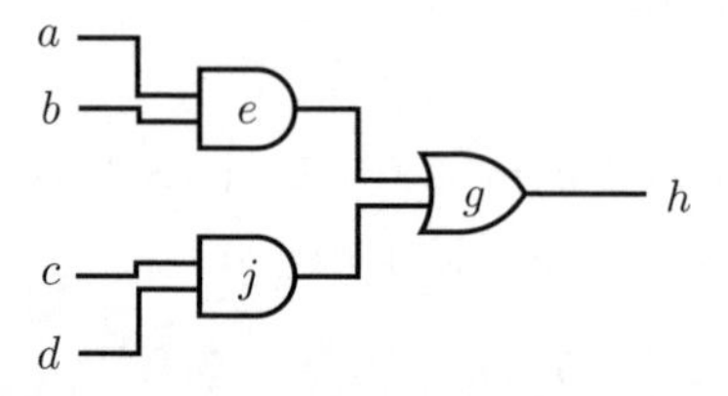

Figure 2.21 Implementation II of $h = ab + cd$

Table 2.3 SPFDs of the gates in Figure 2.21

$SPFD_g^{on}$	$SPFD_g^{off}$	$SPFD_e^{on}$	$SPFD_e^{off}$	$SPFD_j^{on}$	$SPFD_j^{off}$
$ab\bar{c}\bar{d}$	$\bar{a}b\bar{c}\bar{d}$	$ab\bar{c}\bar{d}$	$\bar{a}b\bar{c}\bar{d}$	$\bar{a}\bar{b}cd$	$\bar{a}b\bar{c}\bar{d}$
$ab\bar{c}d$	$\bar{a}\bar{b}\bar{c}\bar{d}$	$ab\bar{c}d$	$\bar{a}\bar{b}\bar{c}\bar{d}$	$a\bar{b}cd$	$\bar{a}\bar{b}\bar{c}\bar{d}$
$abcd$	$a\bar{b}\bar{c}\bar{d}$	$abcd$	$a\bar{b}\bar{c}\bar{d}$	$abcd$	$a\bar{b}\bar{c}\bar{d}$
$abc\bar{d}$	$\bar{a}b\bar{c}d$	$abc\bar{d}$	$\bar{a}b\bar{c}d$	$\bar{a}bcd$	$\bar{a}b\bar{c}d$
$\bar{a}bcd$	$\bar{a}\bar{b}\bar{c}d$		$\bar{a}\bar{b}\bar{c}d$		$\bar{a}\bar{b}\bar{c}d$
$a\bar{b}cd$	$a\bar{b}\bar{c}d$		$a\bar{b}\bar{c}d$		$a\bar{b}\bar{c}d$
$\bar{a}\bar{b}cd$	$\bar{a}bc\bar{d}$		$\bar{a}bc\bar{d}$		$\bar{a}bc\bar{d}$
	$a\bar{b}c\bar{d}$		$a\bar{b}c\bar{d}$		$a\bar{b}c\bar{d}$
	$\bar{a}\bar{b}c\bar{d}$		$\bar{a}\bar{b}c\bar{d}$		$\bar{a}\bar{b}c\bar{d}$

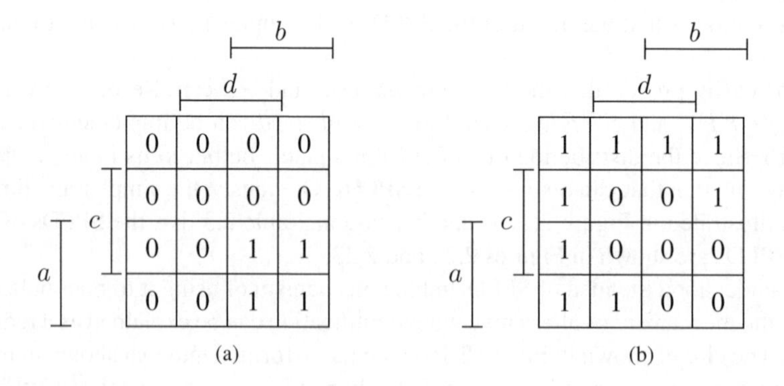

Figure 2.22 New Karnaugh maps of $SPFD_e$. (a) $SPFD_e^{on}$; (b) $SPFD_e^{off}$

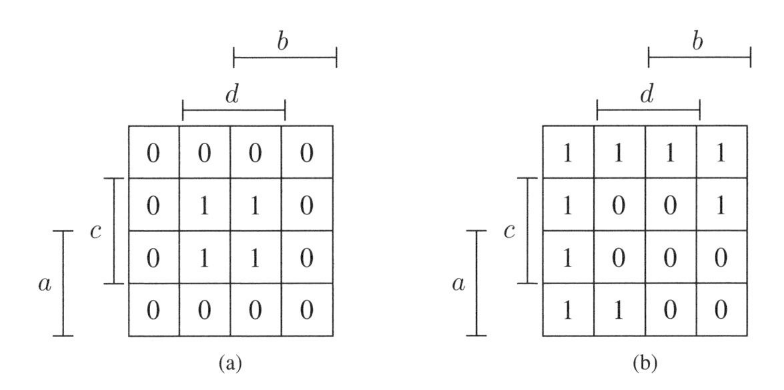

0	0	0	0
0	1	1	0
0	1	1	0
0	0	0	0

1	1	1	1
1	0	0	1
1	0	0	0
1	1	0	0

Figure 2.23 New Karnaugh maps of $SPFD_j$. (a) $SPFD_j^{on}$; (b) $SPFD_j^{off}$

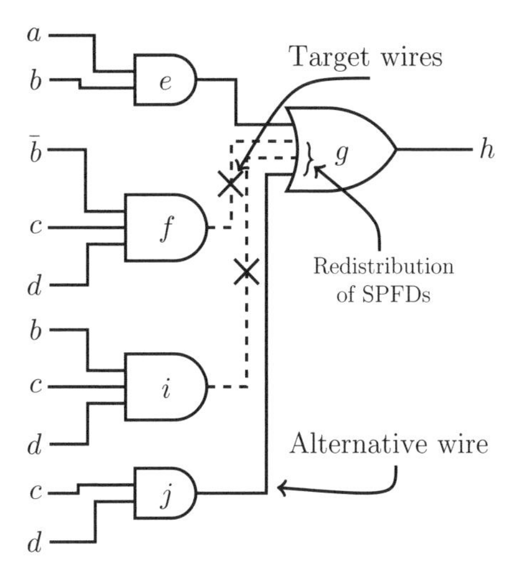

Figure 2.24 SPFD-based rewiring

like error-cancellation-based rewiring techniques. This kind of rewiring techniques is most suitable for programmable devices such as field-programmable gate arrays (FPGAs) where the potential of flexible SPFD redistribution can be fully exploited.

2.4 Why are Rewiring Techniques Important?

Modern devices have reached such a high density that an entire system can be put inside a single chip because of the advances in chip fabrication technology. Transistor size has been scaled down from 45 nm, and then 22 nm, to even below 16 nm in recent years. This allows more efficient chips to be manufactured. However, extremely high silicon density starts to cause new problems in integrated circuit (IC) designs.

First of all, a wire can now be broken easily by an unexpected tiny particle such as dust due to its small size. However, it is difficult to manufacture a chip that is 100% error-free. Another problem is that, at the nanometer scale, the size of a wire is now comparable to that of a gate. Hence, wires have become important and may even be a dominant factor affecting circuit performance in various aspects such as speed, power consumption, and heat dissipation.

Rewiring techniques allows us to remove unwanted wires. This makes rewiring techniques extremely suitable and useful for today's nanometer technologies.

In fact, wire-based circuit optimization techniques are at least as powerful as traditional gate-based circuit optimization techniques. This is because gate-based changes can be modeled by wire-based changes, while the opposite may not necessary be true. For instance, the removal of the wires connecting a gate is equivalent to the removal of the gate. But the removal of a gate cannot specify a particular wire and may involve wires that should not be touched.

A circuit whose gates are fixed and wires are routed is illustrated in Figure 2.25. Gate $g8$ implements the function of the circuit and is the only one primary output. One of inputs of gate $g5$ is hard-wired to the value 1. Among the wires, wire $g2 \rightarrow g4$ is the longest wire. Suppose it causes a timing violation. Since the locations of the gates cannot be changed and it is not possible to perform logic synthesis again, the only method that can solve the timing problem is rerouting the wires. This may not always be feasible, and the router may consider the current result to be the best solution already. In this situation, rewiring techniques may come to the rescue. Wire $g2 \rightarrow g4$ can be regarded as a target wire to be removed. Then, any one of the rewiring techniques can be used to determine the alternative wires of the target

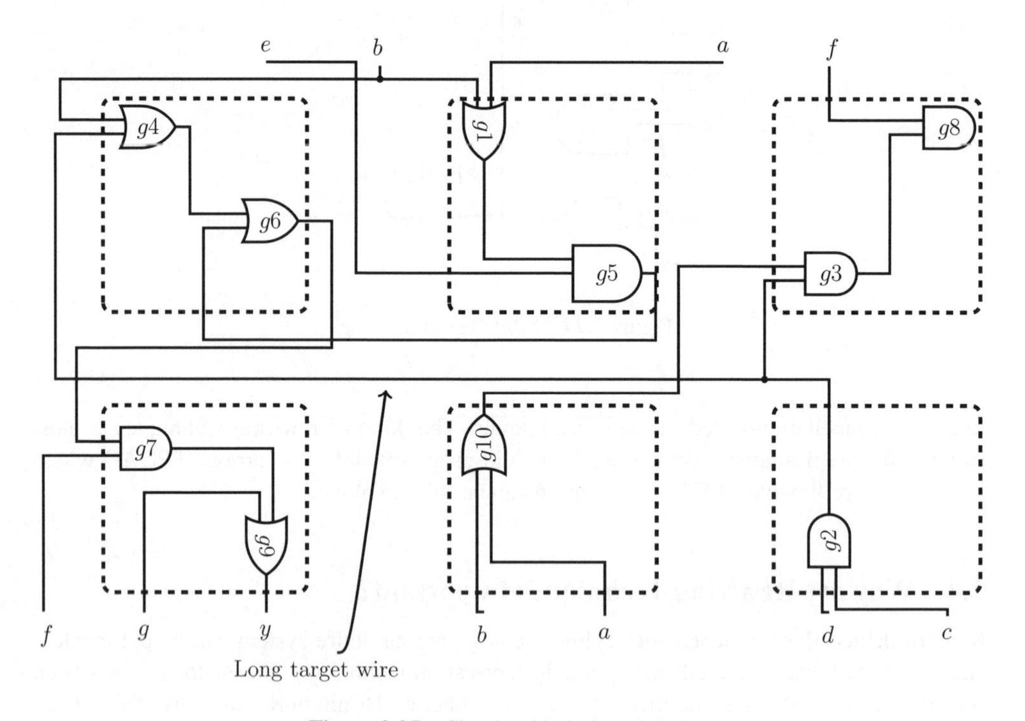

Figure 2.25 Circuit with timing violation

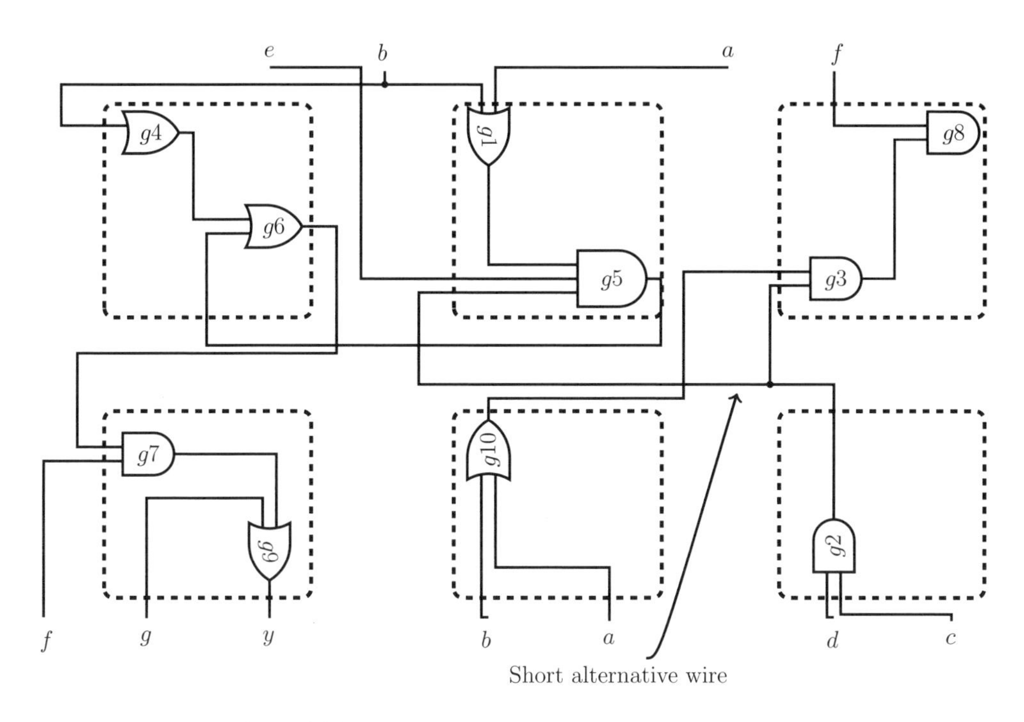

Figure 2.26 Circuit without timing violation after rewiring

wire. It is found that $g2 \rightarrow g5(AND)$ is an alternative wire. Its length is obviously smaller that of wire $g2 \rightarrow g4$. Then the circuit can be transformed into what Figure 2.26 illustrates. No gates and their locations are modified. The only change consists of the modifications made to the hard-wired input of each of the gates $g4$ and $g5$, the removal of wire $g2 \rightarrow g4$, and the addition of wire $g2 \rightarrow g5(AND)$. As a result, if the alternative wire is short enough, the timing violation problem can be solved.

References

S.-C. Chang, L. P. P. P. van Ginneken, and M. Marek-Sadowska. Fast Boolean optimization by rewiring. In *Computer-Aided Design, 1996. ICCAD-96. Digest of Technical Papers, 1996 IEEE/ACM International Conference on*, pages 262–269, November 1996. doi:10.1109/ICCAD.1996.569641.

S.-C. Chang, L. P. P. P. Van Ginneken, and M. Marek-Sadowska. Circuit optimization by rewiring. *IEEE Transactions on Computers*, 48(9):962–970, 1999. ISSN: 0018-9340. doi:10.1109/12.795224.

C.-W. J. Chang and M. Marek-Sadowska. Who are the alternative wires in your neighborhood? (alternative wires identification without search). In Proceedings of the 11th Great Lakes Symposium on VLSI, GLSVLSI '01, pages 103–108, New York, 2001a. ACM. ISBN: 1-58113-351-0. doi:10.1145/368122.368880.

C.-W. Chang and M. Marek-Sadowska. Single-pass redundancy-addition-and-removal. In *Computer Aided Design, 2001. ICCAD 2001. IEEE/ACM International Conference on*, pages 606–609, 2001b. doi:10.1109/ICCAD.2001.968723.

S. C. Chang and M. Marek-Sadowska. Perturb and simplify: multilevel Boolean network optimizer. *IEEE Transactions on Computer-Aided Design of Integrated Circuits and Systems*, 15(12):1494–1504, 1996.

C. W. J. Chang and M. Marek-Sadowska. Theory of wire addition and removal in combinational Boolean networks. *Microelectronic Engineering*, 84(2):229–243, 2007.

C.-W. J. Chang, M.-F. Hsiao, and M. Marek-Sadowska. A new reasoning scheme for efficient redundancy addition and removal. *IEEE Transactions on Computer-Aided Design of Integrated Circuits and Systems*, 22(7):945–951, 2003. ISSN: 0278-0070. doi:10.1109/TCAD.2003.814239.

K.-T. Cheng and L. A. Entrena. Multi-level logic optimization by redundancy addition and removal. In *Design Automation, 1993, with the European Event in ASIC Design. Proceedings. [4th] European Conference on*, pages 373–377, February 1993. doi:10.1109/EDAC.1993.386447.

F. S. Chim, T. K. Lam, and Y.-L. Wu. On improved scheme for digital circuit rewiring and application on further improving FPGA technology mapping. In *Design Automation Conference, 2009. ASP-DAC 2009. Asia and South Pacific*, pages 197–202, 2009. doi:10.1109/ASPDAC.2009.4796480.

F.-S. Chim, T.-K. Lam, Y.-L. Wu, and H. Fan. On structural analysis and efficiency for graph-based rewiring techniques. *IEICE Transactions*, E94(12):2853–2865, 2011.

F. S. Chim and Y.-L. Wu. On extended graph-based rewiring technique. In *ASIC, 2007. ASICON '07. 7th International Conference on*, pages 114–117, 2007. doi:10.1109/ICASIC.2007.4415580.

J. Cong, Y. Lin, and W. Long. SPFD-based global rewiring. In *Proceedings of the 2002 ACM/SIGDA Tenth International Symposium on Field-Programmable Gate Arrays*, FPGA '02, pages 77–84, New York, 2002a. ACM. ISBN: 1-58113-452-5. doi:10.1145/503048.503060.

J. Cong, J. Y. Lin, and W. Long. A new enhanced SPFD rewiring algorithm [logic ic layout]. In *Computer Aided Design, 2002. ICCAD 2002. IEEE/ACM International Conference on*, pages 672–678, November 2002b. doi:10.1109/ICCAD.2002.1167604.

L. A. Entrena and K.-T. Cheng. Combinational and sequential logic optimization by redundancy addition and removal. *IEEE Transactions on Computer-Aided Design of Integrated Circuits and Systems*, 14(7):909–916, 1995. ISSN: 0278-0070. doi:10.1109/43.391740.

M. A. Iyer and M. Abramovici. Fire: a fault-independent combinational redundancy identification algorithm. *IEEE Transactions on Very Large Scale Integration (VLSI) Systems*, 4(2):295–301, 1996. ISSN: 1063-8210. doi:10.1109/92.502203.

R. Krieger, B. Becker, R. Hahn, U. Sparmann. Structure based methods for parallel pattern fault simulation in combinational circuits. In *Proceedings of the European Conference on Design Automation. EDAC.*, pages 497–502. IEEE, 1991.

T.-K. Lam, W.-C. Tang, X. Yang, and Y.-L. Wu. ECR: A powerful and low-complexity error cancellation rewiring scheme. *ACM Transactions on Design Automation of Electronic Systems*, 17(4):50:1 50:21, 2012. ISSN: 1084-4309. doi:10.1145/2348839.2348854.

C.-C. Lin and C.-Y. Wang. Rewiring using IRredundancy removal and addition. In *Design, Automation Test in Europe Conference Exhibition, 2009. DATE '09.*, pages 324 –327, April 2009.

P. Maidee and K. Bazargan. A generalized and unified SPFD-based rewiring technique. In *Field Programmable Logic and Applications, 2007. FPL 2007. International Conference on*, pages 305–310, 2007. doi:10.1109/FPL.2007.4380664.

S. Muroga, Y. Kambayashi, H. C. Lai, and J. N. Culliney. The transduction method-design of logic networks based on permissible functions. *IEEE Transactions on Computers*, 38:1404–1424, 1989. ISSN: 0018-9340. doi:10.1109/12.35836.

S. Sinha and R. K. Brayton. Implementation and use of SPFDs in optimizing Boolean networks. In *Computer-Aided Design, 1998. ICCAD 98. Digest of Technical Papers. 1998 IEEE/ACM International Conference on*, pages 103–110, 1998. doi:10.1109/ICCAD.1998.144252.

X. Wei, T.-K. Lam, X. Yang, W.-C. Tang, Y. Diao, and Y.-L. Wu. Delete and Correct (DaC): an atomic logic operation for removing any unwanted wire. In *VLSI Design*. IEEE Computer Society, 2013.

Y.-L. Wu, W. Long, and H. Fan. A fast graph-based alternative wiring scheme for Boolean networks. *VLSI Design, 2000. Thirteenth International Conference on*, pages 268–273, 2000a. doi:10.1109/ICVD.2000.812620.

Y.-L. Wu, C.-N. Sze, C.-C. Cheung, and H. Fan. On improved graph-based alternative wiring scheme for multi-level logic optimization. *Electronics, Circuits and Systems, 2000. ICECS 2000. The 7th IEEE International Conference on*, vol. 2:654–657, 2000b. doi:10.1109/ICECS.2000.912962.

S. Yamashita, H. Sawada, and A. Nagoya. A new method to express functional permissibilities for LUT based FPGAS and its applications. In *ICCAD '96: Proceedings of the 1996 IEEE/ACM international conference on Computer-aided design*, pages 254–261, Washington, DC, USA, 1996. IEEE Computer Society. ISBN: 0-8186-7597-7.

X. Yang, T.-K. Lam, and Y.-L. Wu. ECR: A low complexity generalized error cancellation rewiring scheme. In *DAC '10: Proceedings of the 47th Design Automation Conference*, pages 511–516, New York, 2010. ACM. ISBN: 978-1-4503-0002-5. doi:10.1145/1837274.1837400.

X. Yang, T.-K. Lam, W.-C. Tang, and Y.-L. Wu. Almost every wire is removable: a modeling and solution for removing any circuit wire. In *Design, Automation Test in Europe Conference Exhibition (DATE), 2012*, pages 1573–1578, 2012. doi:10.1109/DATE.2012.6176723.

3

Add-First and Non-ATPG-Based Rewiring Techniques

3.1 Redundancy Addition and Removal (RAR)

3.1.1 RAMBO

Redundancy addition and removal for multilevel Boolean optimization (RAMBO) (Cheng and Entrena 1993, Entrena and Cheng 1995) is the first ever published rewiring technique. It defined the fundamental mechanism of redundancy addition and removal (RAR). In the process of redundancy addition and removal, a wire w_a that is not in the circuit is searched such that, after its addition, an originally irredundant target wire w_t in the circuit becomes redundant and can be removed. A set of candidate wires whose addition can make w_t redundant is first computed. Then the redundancy of each candidate wire is checked. The function of the circuit is unchanged throughout the process.

Since it is infeasible to try to add wires at every location in the circuit, the authors of RAMBO derived the following lemmas to limit the search space and prune the set of candidate alternative wires: Let g_i and g_j be two logic gates whose outputs are o_i and o_j respectively. Their controlling values are c_i and c_j. The inversion for an NOR, NAND, or NOT gate is defined to be 1, and is defined to be 0 for an OR or AND gate. Let the inversions of g_i and g_j be, respectively, i_i and i_j.

Lemma 3.1 *If the implication $o_i = b_i \Rightarrow o_j = b_j$ holds, where b_i and b_j are Boolean values,*

1. *Suppose g_j is not in the transitive fanin cone of g_i. If $b_i = c_j$ ($\bar{c}_j$) and $b_j = c_j \oplus i_j$, the wire $g_i \rightarrow g_j$ ($g_i \nrightarrow g_j$) is redundant.*
2. *Suppose g_i is not in the transitive fanin cone of g_j. If $b_j = c_i$ ($\bar{c}_i$) and $\bar{b}_i = c_i \oplus i_i$, the wire $g_j \nrightarrow g_i$ ($g_j \rightarrow g_i$) is redundant.*

Boolean Circuit Rewiring: Bridging Logical and Physical Designs, First Edition.
Tak-Kei Lam, Wai-Chung Tang, Xing Wei, Yi Diao and David Yu-Liang Wu.

Lemma 3.2 *Suppose gate g_i is a fanin of gate g_j. Further, suppose g_j has fanins $x_k, k \in 1, \ldots, m$. Let x_i be the output of g_i and is an input of g_j. If $x_i = c_j \Rightarrow x_k = c_j, k = i$, wire $g_i \to g_j$ is redundant.*

The example shown in Figure 2.2(a) in Section 2 shows the above lemmas in action. Lemma 3.1 is used to identify candidate alternative wires to be added to the circuit. On the other hand, Lemma 3.2 is used to identify removable wires. They are applied by assigning a logic value at the output of a gate first. Then, implication based on the assigned value is performed, and the information of redundant target and alternative wires can be obtained. According to the lemmas, more than one alternative wire can be identified in one single implication process. Similarly, more than one wire that are originally irredundant can become redundant. It is safe to add the alternative wires independently. The same applies to the wires that have been made redundant.

3.1.2 REWIRE

The major goal of REWIRE (Chang et al. 1996, Chang and Marek-Sadowska 2001, Chang et al. 1999, Chang and Marek-Sadowska 1996, and Entrena and Cheng 1995) is to further speed up the ATPG-based rewiring scheme.

The concept of mandatory assignment (MA) was studied thoroughly in later techniques of rewiring such as REWIRE. A mandatory assignment is a logic assignment to gates needed for a stuck-at fault test to be performed and must be satisfied by any test vector of the given fault. Observability mandatory assignments (OMAs) are the assignments required for the fault to become observable at primary outputs. Forced mandatory assignments (FMAs) are needed for the fault to be testable. An mandatory assignment is an FMA if using its opposite value in the test fails to test the fault that was originally testable. Based on these definitions, all mandatory assignments obtained by backward propagation are forced.

Several important theorems (Theorems 3.1–3.6) based on the properties of mandatory assignments were proved in Chang et al. (1996) to speed up the rewiring process. These theorems prune the candidate redundant wires after adding alternative wires and help to identify alternative wires. Let C be a circuit, and let w_a be an alternative wire to be added to an AND gate to the circuit. The circuit is assumed to contain no redundant wires.

Theorem 3.1 *All the direct input wires of an AND (OR) gate are not redundant in $C + w_a$ if the AND (OR) gate has an observability and the FMA $=$ 0(1) for the stuck-at fault test of wire w_a.*

Theorem 3.2 *A wire $n_x \to n_z$ is not redundant in $C + w_a$ if n_z is AND (OR) gate and n_x has an observability MA of 1 (0) for the stuck-at fault test of wire w_a.*

Theorem 3.3 *Let node n_z be an AND (OR) gate and one of its input wires is n_x. If n_x has an observability MA of 0 (1) for the stuck-at fault test of wire w_a, all other input wires of n_z are not redundant in $C + w_a$.*

Theorem 3.4 *If wire $w_a = n_s \to n_d$ is an alternative wire for the target wire w_t, node n_s must have a mandatory assignment 0 (1) for the stuck-at fault test of wire w_t and n_d is an AND (OR) gate.*

Theorem 3.5 *If wire $w_a = n_s \rightarrow n_d$ is an alternative wire for the target wire w_t, an AND (OR) gate n_d must have a FMA 1 or D (0 or $\bar{D}$) for the stuck-at fault test of wire w_t.*

Theorem 3.6 *Wire $w_a = n_s \rightarrow n_d$ is an alternative wire for the target wire w_t if and only if an AND (OR) gate n_d has a forced MA = 1 or D (0 $\bar{D}$) and n_s has an MA = 0 (1) for the stuck-at fault test of wire w_t.*

According to the previous discussion and the theorems listed above, it can be concluded that there are two kinds of redundant alternative wires: (i) alternative wires whose addition blocks the fault of the target wire's removal; (ii) alternative wires whose addition invalidates the test of the fault of the target wire's removal. The first kind is constructed at dominators, and the second kind is constructed at nodes with FMAs. We have already shown these two kinds of alternative wires in Figure 2.3. The authors of Chang and Marek-Sadowska (1996) summarized the complete set of feasible alternative wire addition. We reproduce it as Figure 3.1. In the figure, the three-tuple of each type of transformation refers to the mandatory assignment of node g_s, the function of node g_d, and the mandatory assignment (b_g/b_f) of node g_d.

Algorithm 3.1: Identification of alternative wires in REWIRE

```
   input : target wire, w_t
   output: set of alternative wires W_a
1  begin
2      MA ← Collect_MA(w_t) ;
3      foreach gate g_i in the circuit do
4          if g_i has MA then
5              L_src ← L_src ∪ g_i;
6          if g_i is a dominator of w_t or has FMA then
7              L_dst ← L_dst ∪ g_i;
8      foreach g_s ∈ L_src do
9          foreach g_d ∈ L_dst do
10             if g_s has observability MA then
11                 continue;
12             if Is_Redundant (g_s, g_d) then
13                 W_a ← W_a ∪ (g_s → g_d);
14                 (if the value of g_s should inverted, W_a ← W_a ∪ g_s ↛ g_d;
15                 )
16     return W_a;
17 end
```

Since finding all test vectors for a stuck-at fault is hard (an NP-hard problem), the concept of dominators is applied in REWIRE to determine mandatory assignments. Given a nonredundant target wire w_t in the circuit C, stuck-at fault is first performed on the wire to find its alternative wires. The set of mandatory assignments during the fault test is collected. Any gate that has a mandatory assignment is regarded as the source gate of a candidate alternative wire. If a

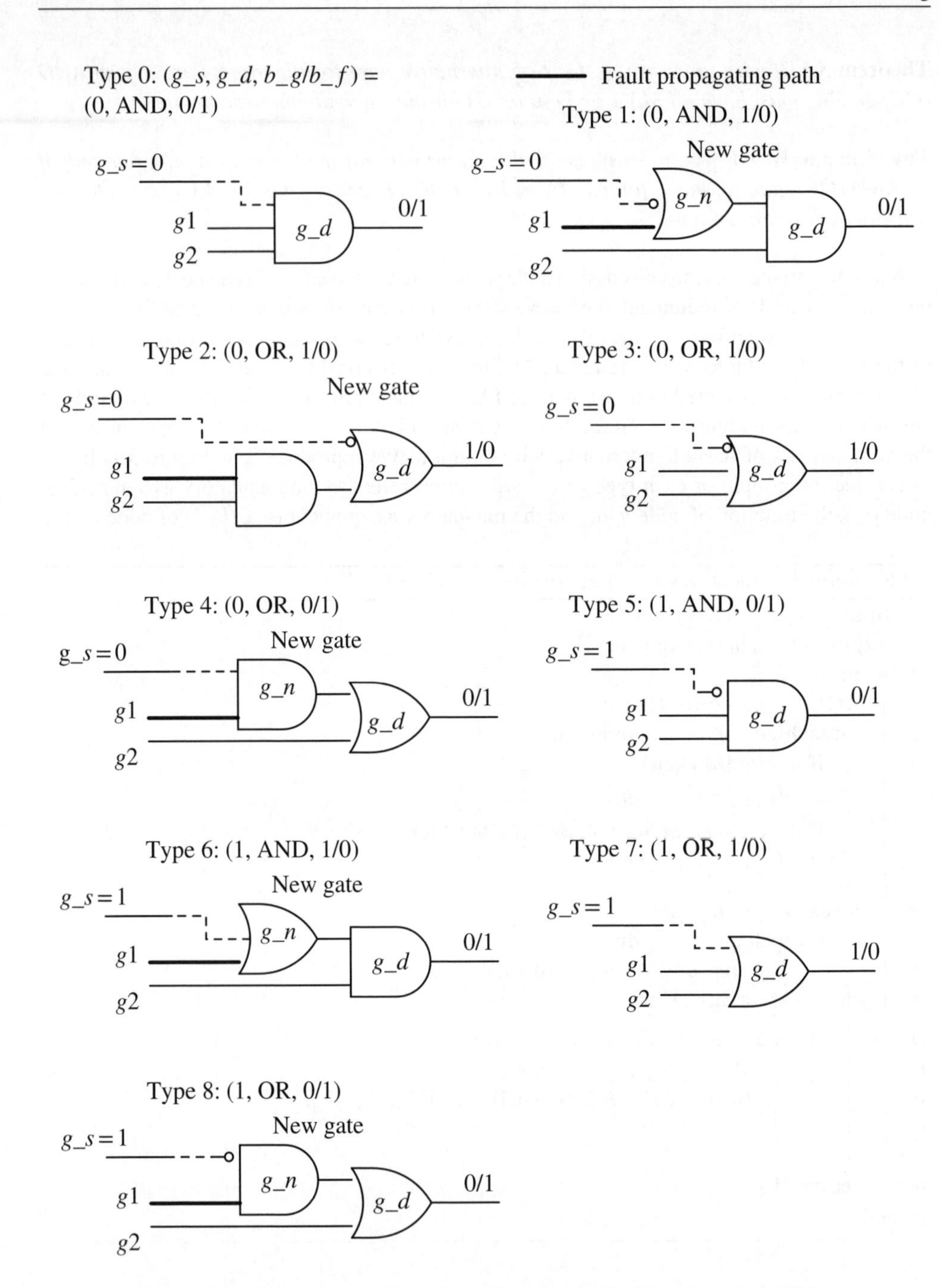

Figure 3.1 Complete set of transformations for single alternative wire addition

gate is the dominator of the target wire, or it has an FMA, it is regarded as the destination gate of a candidate alternative wire. A candidate alternative wire is to be added to the circuit by connecting its source and destination to the inputs of a new gate. The function of the new gate is determined by the mandatory assignments of the source and destination gates. The aim is to block the fault propagation path by creating conflicts at dominators or at gates with FMAs, to make sure that the target wire will be redundant after the addition of the new wire ($C + w_a \equiv C + w_a - w_t$). The candidate transformations, whose additions ensure conflicts in the stuck-at fault test of the target wire, will be checked for redundancy ($C \equiv C + w_a$). A redundant candidate wire will be taken as a valid alternative wire for the target wire. The whole procedure is summarized in Algorithm 3.1.

3.1.3 RAMFIRE

From our previous discussion, it can be understood most RAR-based rewiring techniques consist of two main operations to determine a alternative wire w_a for a target wire w_t. The two main operations are

1. Identifying the candidate alternative wires that do not exist in the circuit whose addition will make the target wire redundant.
2. Determining whether the addition of an alternative wire will change the circuit function. That is, whether the candidate alternative wire is redundant after being added to the circuit.

In RAMFIRE, the redundancy identification technique known as FIRE (Iyer and Abramovici 1996) is applied in the two operations to speed up the rewiring process (Section 2.2.1.3). Recall that in FIRE, we first assign uncontrollability value $\bar{0}$ and then $\bar{1}$ to the target wire w_t and perform implications. The set of gates having uncontrollability or unobservability values assigned as a result of the implication of $w_t = \bar{0}$ is $S(w_t = \bar{0})$, and the set of gates having uncontrollability or unobservability values assigned as a result of the implication of $w_t = \bar{1}$ is $S(w_t = \bar{1})$. The information of the two sets of gates $S(w_t = \bar{0})$ and $S(w_t = \bar{1})$ is then used to determine the redundant wires in the circuit. It is discussed in more detail in this section.

Given a candidate alternative wire w_c whose source gate and destination gate are, respectively, n_{cs} and n_{cd} in a circuit C, and given a set of implication result S, the following notations and theorems are defined:

Definition 3.1 *The source gate n_{cs} is potentially redundant in S if the output of the gate has a mandatory assignment of 0 (1) in the stuck-at fault test of w_t and has a value of $\bar{0}$ ($\bar{1}$) in S.*

Definition 3.2 *The sink gate is n_{cd} is potentially redundant in S if:*

- *the output of n_{cd} is unobservable (assigned with $*$), or*
- *the output of n_{cd} is has a mandatory assignment of 0 (1) in the stuck-at fault test of w_t and has a value of $\bar{0}$ ($\bar{1}$) in S. Moreover, n_{cd} is in the transitive fanout cone of w_t.*

Definition 3.3 *A candidate alternative wire w_c is potentially redundant in S if either its source or destination node is potentially redundant in S.*

Based on the above definitions, the following theorem can be proved:

Theorem 3.7 *A candidate alternative wire w_c is redundant if it is potentially redundant in $S(w_t = \bar{0})$ and $S(w_t = \bar{1})$.*

Proof. Suppose a candidate alternative wire w_c is potentially redundant in both $S(w_t = \bar{0})$ and $S(w_t = \bar{1})$. It is impossible that the source gate is potentially redundant in both $S(w_t = \bar{0})$ and $S(w_t = \bar{1})$. The reason is that in such a case, n_{cs} is in both sets of untestable faults $F(w_t = \bar{0})$ and $F(w_t = \bar{1})$, which implies any wire whose source gate is n_{cs} is redundant in the original circuit. Similarly, it is impossible that the destination gate is potentially redundant in both $S(w_t = \bar{0})$ and $S(w_t = \bar{1})$. Thus, either of the following case is true:

1. The source gate is potentially redundant in $S(w_t = \bar{0})$, and the destination gate is potentially redundant in $S(w_t = \bar{1})$.
2. The source gate is potentially redundant in $S(w_t = \bar{1})$, and the destination gate is potentially redundant in $S(w_t = \bar{0})$.

Assume that the first case is true. Further, assume that n_{cs} has a mandatory assignment of 1 and n_{cd} has a mandatory assignment of 0 in the stuck-at fault test of w_t. The source gate n_{cs} is connected to the destination gate n_{cd} via an OR gate ($n_{cs} \rightarrow n_{cd}$(OR)). The newly added OR gate is denoted by n_{new} (the transformation essentially consists of adding wires: $n_{cs} \rightarrow n_{new}$ and $n_{cd} \rightarrow n_{new}$).

Suppose we assign uncontrollability value $\bar{1}$ to the target wire w_t and perform implication in the modified circuit $C + w_c$. Then, the newly added wire $n_{co} \rightarrow n_{cd}$(OR) is unobservable in the modified circuit. This is because n_{cd} is potentially redundant in $S(w_t = \bar{1})$; and by the definition of the potentially redundant destination node, either of the following cases is true:

(a) The output of n_{cd} in C is replaced by the output of the newly added OR gate n_{new} in $C + w_c$. Since n_{cd} is unobservable in C, the output of n_{new}, and all the inputs of n_{cd} including $n_{cs} \rightarrow n_{cd}$(OR) are also unobservable in $C + w_c$.
(b) $n_{cd} \rightarrow n_{new}$ is implied $\bar{0}$ from its inputs in $C + w_c$, therefore $n_{cs} \rightarrow n_{new}$ is unobservable in $C + w_c$. (Note: If $n_c d$ is in the transitive fanin cone of w_t, its unobservability value $\bar{0}$ is implied from its fanout gates. In this case, the new OR gate n_{new} will block propagation of $\bar{0}$ to n_{cd}. Therefore, we need to ensure n_{cd} is in the transitive fanout cone of w_t.)

In either case, $n_{cs} \rightarrow n_{cd}$(OR) is redundant. Then, suppose we assign uncontrollability value $\bar{1}$ to the target wire w_t and perform implication in the modified circuit $C + w_c$. Source gate n_{cs} will be assigned with uncontrollability value 1. Therefore, the conclusion is that $sa0(w_c)$ is in both $F(w_t = \bar{0})$ and $F(w_t = \bar{1})$ and it is not detectable in $C + w_c$. By similar arguments, other cases can be proved.

The first operation of most RAR-based rewiring techniques, namely identifying candidate alternative wires, is achieved in RAMFIRE by Theorem 3.7. The theorem is also useful and can be modified easily for the second operation.

3.1.4 Comparison Between RAR-Based Rewiring Techniques

It can be seen that REWIRE is an improved version of RAMBO in terms of speed. The major cause is the application of the theorems on OMAs and redundancy. As a result, the number of candidate alternative wires to be evaluated and the time spent on redundancy tests are reduced. RAMFIRE also speeds up REWIRE and RAMBO. This is because the number of ATPG implications are fewer in RAMFIRE due to the utilization of uncontrollability information.

Improvement in an area may come with a cost in other areas. The rewiring ability of REWIRE is similar to that of RAMBO. The information of uncontrollability is much affected by circuit structures and the types of gates used. This is reflected in the rewiring ability of RAMFIRE, which is less powerful than that of RAMBO or REWIRE. The authors of Tang et al. (2003) carried out a quantitative analysis on the rewiring ability of these RAR-based rewiring techniques.

3.2 Node-Based Network Addition and Removal (NAR)

So far, we have discussed the most basic but the most important rewiring concepts and various rewiring techniques. In this section, another kind of logic restructuring technique will be introduced. This kind of techniques is conceptually different from the rewiring concepts discussed previously in such a way that it is node-based instead of wire-based. Their objective is to merge or substitute a node (target node) in a circuit with another existing or new node (alternative node). They can be considered as a kind of rewiring technique to replace more than one wires simultaneously with multiple wires.

In Figure 3.2(a), there is a circuit with two primary outputs $\{o1, o2\}$. It can be proved that gate $g10$ can be removed from the circuit and the function of the circuit can be restored by connecting gate $g8$ to the noninverted upper input of gate $g11$ (without the "bubble"). The result of this feasible transformation is shown in Figure 3.2(b). Hence, gates $g8$ and $g10$ (with inverted output) are said to be mergeable. Since the above operations involve only a removal of a wire and an addition of a wire, traditional rewiring techniques are adequate for the task. However, if $g8$ and/or $g10$ have more than one fanouts, using traditional rewiring techniques may not be suitable.

The development history of node-based transformation techniques is in fact quite similar to that of wire-based ones. First, researchers focused on node merging, in which an existing node in a circuit is replaced by another existing node. Next, it was realized there are some cases where there were no existing nodes suitable to substitute the target nodes. Hence, the approach of adding new nodes as alternative nodes was derived. In this book, node addition and removal (NAR) is used as a collective terminology describing both approaches.

3.2.1 Node Merging

The idea of node merging originated from the researches of formal verification. To be more specific, algorithms were developed to check whether a set of circuits are equivalent. Then, those algorithms have been extended specifically for merging nodes.

Node merging is a logic restructuring method that is commonly used for circuit area reduction. Previously, binary decision diagram (BDD)-based) approaches were popular. However, they cannot scale to large industrial circuits. There are then SAT-based approaches (Zhu et al.

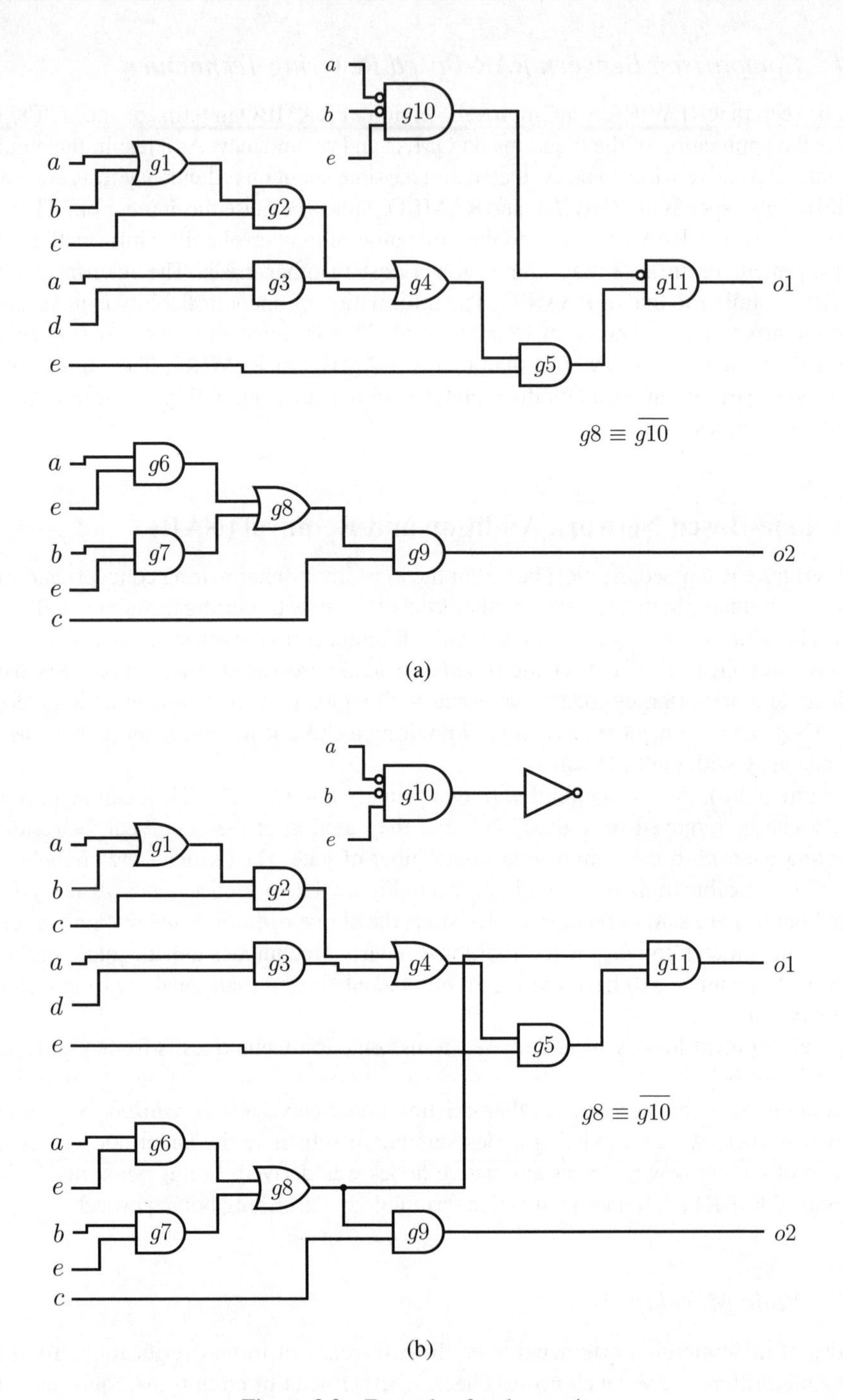

Figure 3.2 Example of node merging

2006, Plaza et al. 2007), and the even more scalable ATPG-based approaches (Chen and Wang 2009, 2010a, 2010b, 2012).

Don't cares are often considered in node-merging algorithms to enhance the number of feasible merges. We briefly describe the method proposed in Plaza et al. (2007) which takes don't cares into account in the following text. The central idea behind this very typical method is based on the definition below, as stated in the publication:

Definition 3.4 *a is ODC-substitutable to b if $ONSET(a) \cup ODC(b) = ONSET(b) \cup ODC(b)$.*

In the definition, $ONSET(a)$ represents the on-set minterms of function a and $ODC(b)$ represents the observability don't cares (ODC) of function b. It means that if don't cares are considered and a is a subset of $b + ODC(b)$, node a can be used to substitute node b. With reference to Figure 3.2, the function of $\overline{g10}$ is $a + b + \bar{e}$. Its ODC is $\bar{e} + \bar{a}\bar{b} + \bar{a}\bar{c} + \bar{d}\bar{c}$. The function of $g8$ is $ae + be$. It is obviously a subset of $ONSET(\overline{g10}) + ODC(g10)$.

It can be realized that this method relies on an efficient ODC computation method. Instead of using symbolic methods, such as BDDs, to calculate ODC, the authors proposed a scalable simulation-based method to calculate approximate ODC.

3.2.1.1 Approximation of Observability Don't Cares

Definition 3.5 *The signature $s(f)$ of a node f in a Boolean circuit is a k-bit sequence of 0s and 1s given by $s(f) = \{f(x_1), f(x_2), \ldots, f(x_{k-1}), f(x_k)\}$, where $x_i, i \in [1..k]$ are input vectors.*

Definition 3.6 *The ODC mask of a node f in a Boolean circuit is a k-bit sequence of 0s and 1s given by $o(f) = \{x_1 \notin ODC_f, x_2 \notin ODC_f, \ldots, x_{k-1} \notin ODC_f, x_k \notin ODC_f\}$, where $x_i, i \in [1..k]$ are input vectors. The corresponding bit position i is assigned 0 if $x_i \in ODC_f$; 1 otherwise.*

The ODC mask of a node is computed from its signature by using the following efficient algorithms. First of all, the circuit is simulated to get the signatures for each of its nodes. The nodes are traversed and processed in the reverse topological order. Primary outputs are fully observable, and all bits of their ODC masks are set to 1. The observability of other nodes are approximated by Algorithm 3.2.

Algorithm 3.2: Appromixation of ODC mask for a node

```
  input : node n
  output:
1 begin
2   foreach f in f ∈ fanouts(n) do
3     local_odc_mask = localOdc(n, f)          // Algorithm 3.3;
4     local_odc_mask = local_odc_mask & odcMask(f);
5     odcMask(n) = odcMask(n) | local_odc_mask;
6 end
```

For every node that is not a primary output, its local ODC mask with respect to each of its fanouts is calculated. Trace bits in its signature are flipped to see whether there are error effects showing up at the fanout (Algorithm 3.3). The local ODC masks obtained are then processed to take care of the fanout's ODC and the multi-fanout case.

Algorithm 3.3: Calculation of local ODC mask for a node with respect to one fanout

input : node n, node fanout f
output: local odc mask of n with respect to f

1 **begin**
2   trace_value = `signature`(f);
3   new_value = `gateFunction`(f, `bitsFlip`(`signature`(n)));
4   local_mask = trace_value $\oplus$ new_value;
5   **return** *local_mask*;
6 **end**

Figure 3.3 depicts an example circuit showing the signatures (s) and the ODC masks (o) for each of the nodes, which are calculated by the approximation algorithms introduced.

The accuracy of this ODC approximation method was addressed. It was reported that there were at most 2.8% incorrect non-ODCs. And at most 9% of ODCs were approximated as non-ODCs. Since this method was applied only as an evaluation metric in the paper, the inaccuracy was acceptable as a tradeoff in view of the advantages provided.

3.2.1.2 Identifying ODC-Based Node Mergers

Once the ODCs of a node n have been calculated, the "lower bound" and "upper bound" of the node's function can be obtained. The lower bound is defined as $ONSET(n) \cdot \overline{ODC(n)}$, and the upper bound is defined as $ONSET(n) + ODC(n)$. If there is a node m whose function is equivalent to the lower bound of node n, or is a subset of the upper bound of node n, it can be used to substitute node n. How can such alternative nodes be determined?

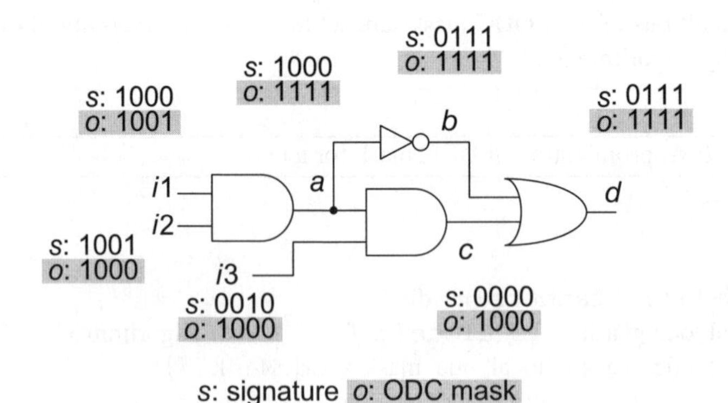

Figure 3.3 Signatures and ODC masks

Since a circuit has been simulated to compute the approximate ODCs, we can make use of the simulation data to identify candidate alternative nodes for a node quite efficiently. The trick is to treat the simulation data as numbers. The lower and upper bound of a node's function can be calculated via bitwise operations. The problem of determining whether a node's function is between the lower and upper bound of another node can be modeled as a problem of checking whether a number is within a certain range. For example, in Figure 3.3, the values for different test vectors of node $i1$ in the circuit are 1 followed by three zeroes. They can be viewed as the binary number 1000. Its lower and upper bound are $1000\&1001 = 1000$ and $1000|0110 = 1110$, respectively.

Then, finding candidate alternative nodes can be achieved by a series of binary searches on the sorted signatures of all nodes. It is because as the ODCs are approximated, formal verification is necessary to check whether a candidate alternative node is valid. SAT-based verification was adopted by the authors of Plaza et al. (2007) to verify the correctness of candidate alternative nodes. They designed an efficient SAT-based verification scheme. Interested readers may refer to their paper for further details.

3.2.1.3 From SAT-Based to ATPG-Based

The node-merging approach that has just been introduced aims to decrease the complexity of computing ODC. The authors of Chen and Wang (2009, 2010a) further improved node merging. They proposed not to apply SAT for verification, and in fact, not to rely on verification at all. Instead, their method makes use of ATPG theories to get rid of trials and errors. Don't cares are taken into account implicitly.

The process of replacing a node by another node is modeled as a process of error testing. If no test vectors can activate and propagate the error induced by the change, the replacement is valid. The original circuit is denoted by C, and the circuit after node merging is denoted by C'. Theorem 3.8 was introduced in Chen and Wang (2009) to state the conditions of a test vector for detecting a node replacement error.

Theorem 3.8 *Let f denote an error of replacing n_t with n_s. An input vector t in C will be a test for f if and only if t generates different values for n_t and n_s, and the effect of the value of n_t is observable at a PO.*

Proof. If an input vector t generates different values for n_t and n_s, and the effect of the value of n_t is observable at a PO, the functions of n_t and n_s can be determined to be different. Then, vector t is a test vector for f. It is obvious that if t is a test vector for f, it generates different values for n_t and n_s and the difference is observable. ∎

The relationship between the test vectors and a node replacement error was analyzed. The test vectors for the error of replacing n_t with n_s are classified into two subsets:

1. $t \in T_0$ if and only if $n_t = 0, n_s = 1$,
2. $t \in T_1$ if and only if $n_t = 1, n_s = 0$.

Under conditions where t is test vector for the node replacement error

1. if $t \in T_0$, it is a test vector for $sa1(n_t)$,
2. if $t \in T_1$, it is a test vector for $sa0(n_t)$.

An error of node replacement f can be divided into two subsets:

1. $f \in f_1$ if and only if $n_t = 0, n_s = 1$,
2. $f \in f_0$ if and only if $n_t = 1, n_s = 0$.

From the definitions, it can be seen that T_0 is the set of test vectors for testing f_1, and T_1 is the set of test vectors for testing f_0. If both T_0 and T_1 are empty, both errors f_1 and f_0 are untestable and thus error f is untestable. Since $n_s = 1$ and $n_s = 0$ are part of the necessary conditions for detecting errors f_1 and f_0, respectively, if n_s cannot be set to 1 and 0 in the corresponding test, error f cannot be detected.

Based on this analysis, the authors of Chen and Wang (2009) derived a sufficient condition (Condition 3.1) to identify the alternative nodes of a given target node.

Condition 3.1 *Let f denote the error of replacing n_t with n_s. If there are conflicts in mandatory assignments: $n_s = 0$ during the stuck-at fault test for $sa1(n_t)$ and $n_s = 1$ during the stuck-at fault test for $sa0(n_t)$, f cannot be detected.*

Here is the procedure to find the alternative nodes $n_s \in N_s$ for a given target node n_t. First, tests for both $sa0(n_t)$ and $sa1(n_t)$ are conducted. Then, nodes with mandatory assignments satisfying (Condition 3.1) can be identified. If a member in this set of nodes is not in the transitive fanout of node n_t, it is added to the set of alternative nodes N_s. This requirement ensures that there will not be cyclic combinational circuits. Since it is possible to add inverters at the output of the alternative nodes, the overall algorithm of finding alternative nodes is as listed in Algorithm 3.4.

Algorithm 3.4: Procedure *Find_Alternative_Nodes*

input : Node n_t
output: set of alternative nodes N_s
1 **begin**
2 $T_0 \leftarrow$ Compute MAs of $sa1(n_t)$;
3 $T_1 \leftarrow$ Compute MAs of $sa0(n_t)$;
4 $N_s \leftarrow$ nodes having different values in T_0 and T_1 and not in the transitive fanout cone of n_t;
5 **return** N_s;
6 **end**

3.2.2 Node Addition and Removal

It is not always possible to obtain alternative nodes for a target node in a circuit. The authors of Chen and Wang (2009) further extended their work to a more powerful logic restructuring technique named node addition and removal (NAR) (Chen and Wang 2010a, 2010b, 2012). In this improved version of the node-merging technique, additional nodes can be added to a circuit to replace a node.

Condition 3.1 may be applied to decide whether a newly added node can replace a target node, but it is not suitable. This is because there are too many possible locations to add a node, and it is inefficient to evaluate all additions one by one. Therefore, the authors proposed converting the problem into finding two existing nodes in the circuit that can form a new node to replace the target node. As a result, two new conditions were derived.

Let $imp(A)$ be the set of value assignments that can be logically implied from a set of value assignment A. Let n_t be the target node and n_a be the additional node. Suppose node n_a is driven by n_{f1} and n_{f2} directly. The sets of mandatory assignments for the tests of $sa0(n_t)$ and $sa1(n_t)$ are, Respectively, denoted by $MAs(n_t = sa0)$ and $MAs(n_t = sa1)$. Boolean circuits are represented as and-inverter graphs.

Condition 3.2 *If both $n_{f1} = 1$ and $n_{f2} = 1$ are the mandatory assignments for the test of $sa0(n_t)$, $n_a = 1$ is also an mandatory assignment for the test.*

Condition 3.3 *If $n_{f2} = 0$ is a value assignment in $imp((n_{f1} = 1) \cup MAs(n_t = sa1))$, $n_a = 0$ is also a mandatory assignment for $sa0(n_t)$.*

For Condition 3.2, it is because as $n_a \equiv n_{f1} \wedge n_{f2}$, $n_a = 1$ is the logical implication of $n_{f1} = 1$ and $n_{f2} = 1$. With reference to Condition 3.1, the latter half of the requirement is fulfilled if Condition 3.2 is satisfied. Similarly, if Condition 3.3 is satisfied, the first half of Condition 3.1 is fulfilled. Hence, when both Conditions 3.2 and 3.3 are satisfied, n_a can be used to replace n_t.

Algorithm 3.5: Procedure *Find_Added_Alternative_Nodes*

input : Node n_t
output: set of alternative nodes N_a
1 **begin**
2   Compute $MAs(n_t = sa0)$;
3   Compute $MAs(n_t = sa1)$;
4   **foreach** *MA* $n_{f1} = v$ *in* $MAs(n_t = sa0)$ **do**
5   Compute $imp((n_{f1} = v) \cup MAs(n_t = sa1))$;
6     $n_{f2} \leftarrow$ nodes whose value assignments are different in $MAs(n_t = sa0)$ and $imp((n_{f1} = v) \cup MAs(n_t = sa1))$;
7   $N_a \leftarrow N_a \cup$ nodes formed by fanins n_{f1} and n_{f2} ;
8   **foreach** *MA* $n_{f1} = v$ *in* $MAs(n_t = sa1)$ **do**
9   Compute $imp((n_{f1} = v) \cup MAs(n_t = sa0))$;
10     $n_{f2} \leftarrow$ nodes whose value assignments are different in $MAs(n_t = sa1)$ and $imp((n_{f1} = v) \cup MAs(n_t = sa0))$;
11   $N_a \leftarrow N_a \cup$ nodes formed by fanins n_{f1} and n_{f2} ;
12   **return** N_a;
13 **end**

It has been assumed that node n_a is driven directly by n_{f1} and n_{f2}. In fact, the output of each of the nodes n_{f1}, n_{f2}, and n_a can be complemented. Hence, Conditions 3.2 and 3.3 can

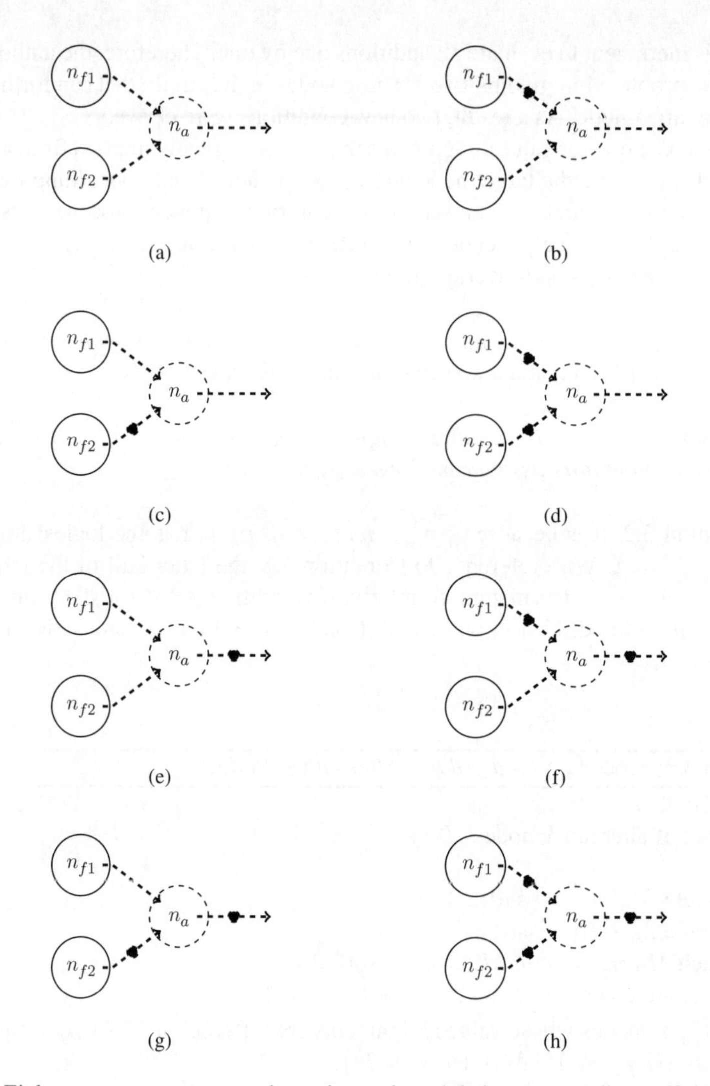

Figure 3.4 Eight ways to construct an alternative node and their corresponding sufficient conditions.
(a) $\{n_{f1} = 1, n_{f2} = 1\} \in MAs(n_t = sa0)n_{f2} = 0 \in imp((n_{f1} = 1) \cup MAs(n_t = sa1))$;
(b) $\{n_{f1} = 0, n_{f2} = 1\} \in MAs(n_t = sa0)n_{f2} = 0 \in imp((n_{f1} = 0) \cup MAs(n_t = sa1))$;
(c) $\{n_{f1} = 1, n_{f2} = 0\} \in MAs(n_t = sa0)n_{f2} = 1 \in imp((n_{f1} = 1) \cup MAs(n_t = sa1))$;
(d) $\{n_{f1} = 0, n_{f2} = 0\} \in MAs(n_t = sa0)n_{f2} = 1 \in imp((n_{f1} = 0) \cup MAs(n_t = sa1))$;
(e) $\{n_{f1} = 1, n_{f2} = 1\} \in MAs(n_t = sa1)n_{f2} = 0 \in imp((n_{f1} = 1) \cup MAs(n_t = sa0))$;
(f) $\{n_{f1} = 0, n_{f2} = 1\} \in MAs(n_t = sa1)n_{f2} = 0 \in imp((n_{f1} = 0) \cup MAs(n_t = sa0))$;
(g) $\{n_{f1} = 1, n_{f2} = 0\} \in MAs(n_t = sa1)n_{f2} = 1 \in imp((n_{f1} = 1) \cup MAs(n_t = sa0))$;
(h) $\{n_{f1} = 0, n_{f2} = 0\} \in MAs(n_t = sa1)n_{f2} = 1 \in imp((n_{f1} = 0) \cup MAs(n_t = sa0))$

be generalized into eight cases. They are shown in Figure 3.4. A black circle on an edge means that the value of the source is complemented. Algorithm 3.5 lists the general algorithm of node addition and removal.

3.3 Other Rewiring Techniques

3.3.1 SPFD-Based Rewiring

The existence of functional flexibilities at different points in a logic circuit is a necessity for rewiring. While ATPG-based optimization methods implicitly use functional flexibilities through activations and propagations of stuck-at fault effects, other optimization methods express flexibilities explicitly. We name such flexibilities as *don't cares*(symbol: $\times$). Functions expressed with don't cares are said to be *incompletely specified.* In previous chapters, we have seen the usage of satisifiability don't cares (SDC) and ODC in rewiring. However, to more formally express and represent functional flexibilities, *set of pairs of functions to be distinguished* (SPFD) is preferred, as it has been shown to be more powerful than its conventional counterparts.

The use of SPFDs in rewiring is strongly tied with circuits implemented with field-programmable gate arrays (FPGAs), where logic nodes are usually implemented by lookup tables (LUTs) and are more complex than simple logic gates.

3.3.1.1 Definitions of SPFDs

Definition 3.7 *Given two logic functions f and g, if for all input vectors X,*

$$g(X) = 1 \Rightarrow f(X) = 1,$$

then f is said to **include** *g, and we write $g \leq f$ or $g \Rightarrow f$.*

It follows that f includes g if and only if $g \cdot \bar{f} \equiv 0$, and we can verify the inclusion using a tautology check of the product of the functions. This concept of function inclusion can also be interpreted as ON-set inclusion: f includes g ($g \leq f$) when the ON-set of g is a subset of that of f.

Definition 3.8 *A function f is said to* **distinguish** *a pair of functions g_1 and g_2 if either of the following conditions is satisfied:*

- *f includes g_1, and $\bar{f}$ includes g_2 ($g_1 \leq f \leq g_2$);*
- *f includes g_2, and $\bar{f}$ includes g_1 ($g_2 \leq f \leq g_1$).*

In other words, f distinguishes g_1 and g_2 when one of its ON-set and OFF-set includes g_1 and the other includes g_2. For example, the function $f = (x_3 + x_2)x_1$ distinguishes the pair ($g_1 = x_2x_1, g_2 = \bar{x_1}$) since f includes g_1 ($g_1 \cdot \bar{f} = (\bar{x_3} \cdot \bar{x_2} + \bar{x_1})(x_2x_1) = 0$), and $\bar{f}$ includes g_2 ($g_2 \cdot f = 0$). The functions are also visualized using Karnaugh maps (K-maps) in Figure 3.5. It should be noted that g_1 and g_2 are mutually exclusive and contains no common ON-points, that is, $g_1 \cdot g_2 \equiv 0$.

Therefore we can collect a pair of functions in a set and call that an **SPFD**. We are mainly interested if some function f can distinguish every pair in the set.

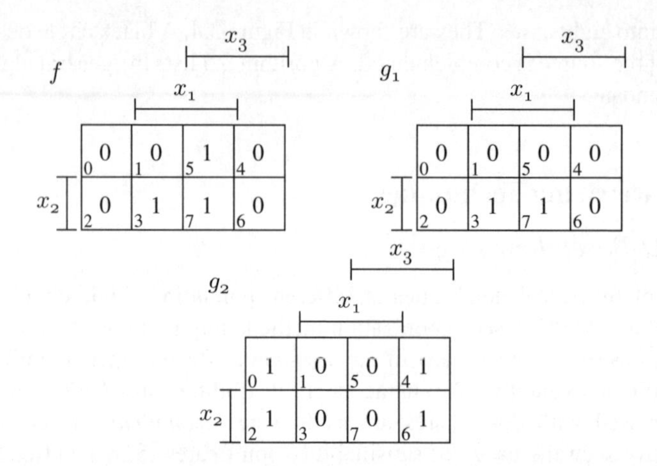

Figure 3.5 SPFD: Example 1

Definition 3.9 *Given an SPFD* $\mathbf{S} = \{(g_{1a}, g_{1b}), (g_{2a}, g_{2b}), \ldots, (g_{na}, g_{nb})\}$, *a function* f *is said to* ***satisfy*** *the SPFD condition (or simply satisfy the SPFD) if* f *distinguishes all* g_{ia} *and* g_{ib}.

For example, the function $f = \bar{x_2} + x_1$ satisfies the SPFD $\mathbf{S} = \{(\bar{x_3}x_2\bar{x_1}, \bar{x_3}\bar{x_2}\bar{x_1}), (\bar{x_3}\bar{x_2}x_1 + x_3x_2x_1, x_3\bar{x_2}\bar{x_1})\}$, as shown in Figure 3.6. Actually there are more functions that satisfy the SPFD, and in general these functions can be expressed as the K-map in Figure 3.6. It should be noted that when there are more than one function pair in an SPFD, then the SPFD cannot

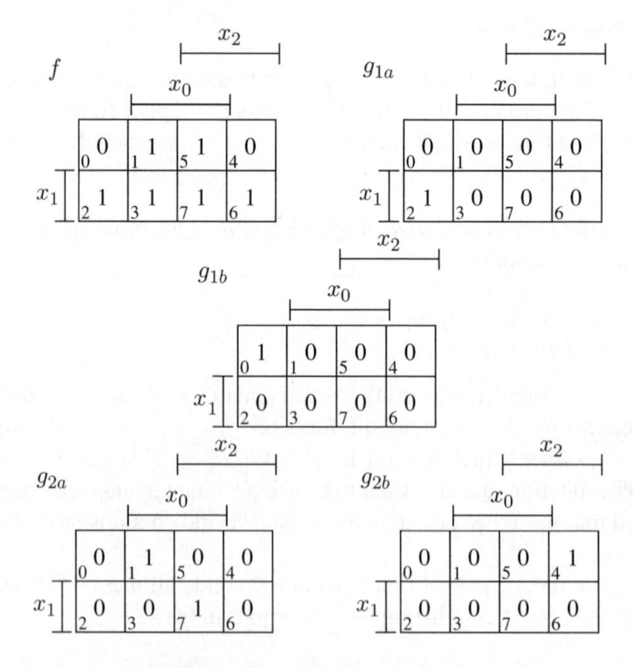

Figure 3.6 SPFD: Example 2

be expressed as an incompletely specified function. In other words, SPFD can represent more functional flexibilities than its counterparts.

3.3.1.2 Calculation of SPFDs

Suppose we are given a Boolean network, and we are interested to calculate SPFDs at each node and each fanout branch. These SPFDs are generally stored for later retrieval when we proceed to the rewiring process. In the following, we will represent a node as N_i and a wire as a pair of nodes (N_i, N_j), and so the SPFDs stored at the node and the wire are represented by SPFD$[N_i]$ and SPFD$[(N_i, N_j)]$, respectively.

The whole calculation procedure begins with initialization: for each node N_i, its function f_i is determined, and a table for recording the SPFD for each node is made empty. Then the recursive procedure *Cal_SPFD* will be invoked for all primary input nodes. The procedure is summarized in Algorithm 3.6. This procedure is responsible for calculating the SPFD for a given node N_i and then propagating it to the input wires of N_i (to be discussed later). There are two trivial cases: (i) the SPFD has been calculated previously; (ii) the node is a primary output. For case (i), we just simply return the calculated SPFD, while in case (ii) we can generate a trivial SPFD containing the pair $(f_i, \bar{f}_i)$. Otherwise, we need to aggregate the function pairs from all fanouts of N_i. Therefore, the procedure is recursively called on the fanouts. Suppose we let SPFD$[(N_i, N_j)]$ be $\{(g_{[i,j]1a}, g_{[i,j]1b}), \ldots, (g_{[i,j]ka}, g_{[i,j]kb}), \ldots\}$, then f_i always satisfies the SPFD(s) and so f_i includes $g_{[i,j]ka}$ and $\bar{f}_i$ includes $g_{[i,j]kb}$, $\forall k$. We may simplify the SPFD for N_i to $\{(f_{ON}, f_{OFF})\}$, where f_{ON} and f_{OFF} are the Boolean sum of all first(second) functions from all pairs and all fanouts, respectively. Leaving only one pair in the node SPFD

Algorithm 3.6: Procedure *Cal_SPFD*

```
   input : node N_i
   output: SPFD S
1  begin
2  |  if SPFD[N_i] ≠ φ then
3  |  |  return SPFD[N_i];
4  |  else if N_i is a primary output then
5  |  |  return {(f_i, f̄_i)};
6  |  else
7  |  |  foreach fanout N_j of N_i do
8  |  |  |  SPFD[N_j] ← Cal_SPFD(N_j);
9  |  Let SPFD[(N_i, N_j)] be{(g_[i,j]1a, g_[i,j]1b), ..., (g_[i,j]ka, g_[i,j]kb), ...}
   |
   |        S ← { ( Σ_{∀j,k} g_[i,j]ka , Σ_{∀j,k} g_[i,j]kb ) }
   |
10 |  if N_i is not a primary input then
11 |  |  Propagate_SPFD(N_i);
12 end
```

$$\mathbf{S} \leftarrow \left\{ \left(\sum_{\forall j,k} g_{[i,j]ka}, \sum_{\forall j,k} g_{[i,j]kb} \right) \right\}$$

probably makes the procedure more efficient. It should be noted that there are more than one pairs in SPFDs of wires.

Once the SPFD for the node N_i is calculated, we have to propagate that to the fanin side. N_i can be complex and has p fanin nodes, which are represented by $N_1, N_2, \ldots, N_p$. The main idea is to distribute the function pairs in the SPFD to that of the fanins by verifying the satisfiability of the functions at the fanins. And this propagation of SPFD is mainly divided into three main subroutines called *Propagate_SPFD, Divide_SPFD*, and *Assign_SPFD*, which will be detailed in the following.

The *Propagate_SPFD* subroutine (see Algorithm 3.7) initializes the SPFDs of the fanin wires of node N_i and assigns function pairs into them. This is the main driver for *Divide_SPFD* and *Assign_SPFD*.

Algorithm 3.7: Procedure *Propagate_SPFD*

```
  input: node N_i
1 begin
2   foreach fanin node N_k of N_i do
3     SPFD[(N_k, N_i)] ← φ;
4   F ← Divide_SPFD(N_i);
5   Assign_SPFD(F, N_i);
6 end
```

The SPFD of a node N_i contains only a function pair (f_{ON}, f_{OFF}) and the Boolean sum of the functions will give the care-set of the SPFD. Therefore, we can generate the set of 2^p minimum product terms (relative to this node) and multiply them with the care-set, before partitioning them into either $\mathbf{F_1}$ or $\mathbf{F_0}$.

A *minimum product term* (MP-term) is defined to be the local function product of all fanin functions. For instance, a node N with p fanins whose functions $g_1, g_2, \ldots, g_p$ has 2^p MP-terms, namely

$$b_0 = \bar{g}_1\bar{g}_2 \cdots \bar{g}_n$$

$$b_1 = \bar{g}_1\bar{g}_2 \cdots g_n$$

$$\vdots$$

$$b_{2^p} = g_1 g_2 \cdots g_n$$

A MP-term is *restricted* when it is multiplied with the care-set of a function pair (f_{ON}, f_{OFF}):

$$a_i = (f_{ON} + f_{OFF}) \cdot b_i$$

$\mathbf{F_1}(\mathbf{F_0})$ simply contains the restricted MP-terms included by $f_{ON}(f_{OFF})$. The Cartesian product of the two sets gives the SPFD. These pairs are named as *atomic* function pairs, which are now ready for assignment to a certain fanin of N_i. The procedure is illustrated in Algorithm 3.8.

Given an atomic pair (a_l, a_m), it is not difficult to find function(s) to distinguish the pair: compare bits of l and m and locate the position where the bit of l is the opposite of that of m (this must exist since $l \neq m$). For instance, f_2 can distinguish the pair $(a_2, a_3) = (a_{010}, a_{011})$. When there are more than one locations, then we can choose any one of them. This leaves

Algorithm 3.8: Procedure *Divide_SPFD*

```
  input  : node N_i with p fanins
  output: SPFD F
1 begin
      /* Generate MP-terms b_i, b_0 = b_000 = f̄_1 f̄_2 f̄_3, etc          */
2     B ← {b_0, b_1, ..., b_i, ..., b_2P};
3     foreach b_i ∈ B do
          /* Generate restricted MP-terms                              */
4         a_i = b_i · (f_ON + f_OFF);
5     F_1 ← {a_i | nonempty a_i included by f_ON};
6     F_0 ← {a_i | nonempty a_i included by f_OFF};
7     return (F_1 × F_0);
8 end
```

room for the heuristic to make the decision. An easy way is to choose the smallest position as k, so the pair will be assigned to the kth fanin of the node N_i. The operator Δ in Algorithm 3.9 abstracts the method to pick an appropriate position.

Algorithm 3.9: Procedure *Assign_SPFD*

```
  input: node N_i, SPFD F
1 begin
2     foreach pair (a_l, a_m) ∈ F do
3         k ← Δ(l, m);
4         SPFD[(N_k, N_i)] ← SPFD[(N_k, N_i)] ∪ {(a_l, a_m)};
5 end
```

We will explain clearly the following example to illustrate the whole procedure to propagate a node SPFD to its fanins. Suppose we have a node N_4 having three fanins N_1, N_2, and N_3, and the SPFD for N_4 is given as $\{(f_{ON}, f_{OFF})\} = \{(x_3x_1 + \bar{x}_3x_2, \bar{x}_2\bar{x}_1)\}$. The functions of N_1, N_2, and N_3 are f_1, f_2, and f_3, respectively. The corresponding K-maps are as follows:

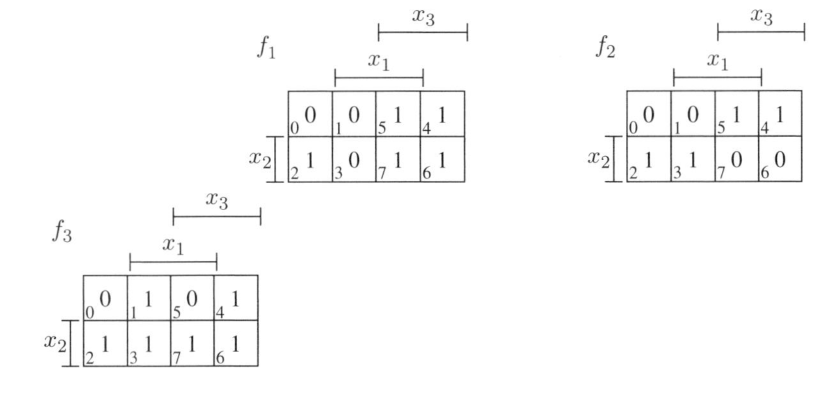

In the procedure *Divide_SPFD*, we generate all eight MP-terms and the corresponding restricted MP-terms. It results in six nonzero restricted MP-terms, which are separated into $\mathbf{F}_1$ and $\mathbf{F}_0$, respectively (see below). The Cartesian product $\mathbf{F}$ contains the function pairs to be distinguished.

$$\mathbf{F}_1 = \{a_{001}, a_{111}\}$$
$$\mathbf{F}_0 = \{a_{000}, a_{011}, a_{101}, a_{110}\}$$
$$\mathbf{F} = \mathbf{F}_1 \times \mathbf{F}_0$$
$$= \left\{ \begin{array}{llll} (a_{001}, a_{000}), & (a_{001}, a_{011}), & (a_{001}, a_{101}), & (a_{001}, a_{110}), \\ (a_{111}, a_{000}), & (a_{111}, a_{011}), & (a_{111}, a_{101}), & (a_{111}, a_{110}) \end{array} \right\}$$

The next step is to assign the pairs of the SPFD to an appropriate fanin. Suppose the leftmost different bit method is used, then we will have the following assignment:

$$\text{SPFD}[(N_1, N_4)] = \{(a_{001}, a_{101}), (a_{001}, a_{110}), (a_{111}, a_{000}), (a_{111}, a_{011})\}$$
$$\text{SPFD}[(N_2, N_4)] = \{(a_{001}, a_{011}), (a_{111}, a_{101})\}$$
$$\text{SPFD}[(N_3, N_4)] = \{(a_{001}, a_{000}), (a_{111}, a_{110})\}$$

And there are 32 possible functions that satisfy SPFD$[(N_3, N_4)]$, as expressed by the K-map below.

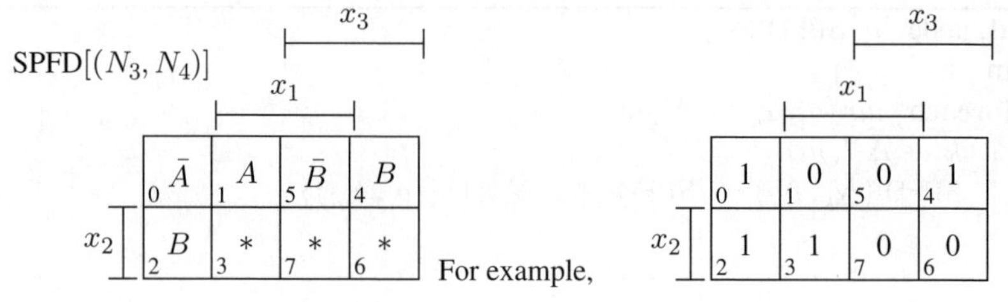

3.3.1.3 SPFD Local Rewiring (SPFD-LR)

It is obvious that when a node N_a with function f_a satisfies the SPFD of target wire (say SPFD$[(N_k, N_i)]$), then it is valid that we can replace the target wire with an alternative wire (N_a, N_i). But it is necessary to modify the function of N_i (i.e., f_i) so that SPFD$[N_i]$ is still satisfied. Since the target wire and the alternative wire share the same destination node N_i, this basic SPFD-based rewiring scheme is named *local rewiring*, and will be abbreviated as *SPFD-LR* in this text. Figure 3.7 depicts the basic structure used by SPFD-LR.

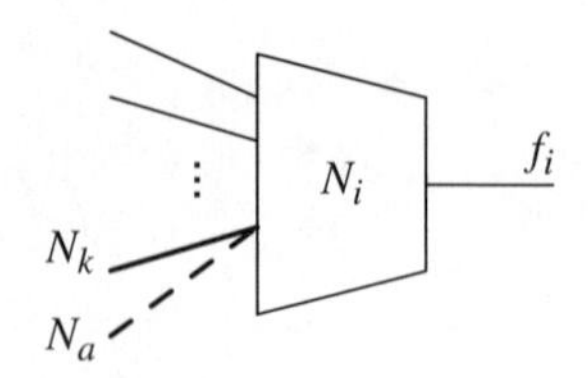

Figure 3.7 SPFD-based local rewiring

The next central question is how to modify the internal logic of N_i and how to ensure that the modification is correct to preserve the satisfiability of the original SPFD. First we will study the procedure to modify the internal logic. The main idea is to check the inclusion of the new set of fanin functions with the atomic function pairs from sets $\mathbf{F_1}$ and $\mathbf{F_0}$, as defined in *Divide_SPFD*. The whole procedure *LR_Modify_Logic* is summarized in Algorithm 3.10. The internal logic of node N_i is expressed as a sum of minterms m_r of the fanin variables $y_1, y_2, \ldots, y_n$. At first, all minterms are initialized to logic 1. The fanin variables y_i will be multiplied to the minterms in the phase decision by the inclusion of a_l by f_k. As an example, consider a target wire (N_3, N_4) as depicted in the previous example, and suppose we have a node N_a with function $f = \bar{x}_2 + x_1$ in Figure 3.6. Clearly f satisfies SPFD$[(N_3, N_4)]$, and therefore we can locally replace the target wire with an alternative wire (N_a, N_4), with a proper modification of the function f_4.

Algorithm 3.10: Procedure *LR_Modify_Logic*

```
    input: node N_i
1   begin
        /* enumerate functions in F_1 and F_0                    */
2       for r from 1 to |F_1| do
3           a_l ← rth function in F_1;
4           for j from 1 to |F_0| do
5               a_m ← jth function in F_0;
6               k ← Δ(l, m);
7               if f_k includes a_l then
8                   m_r ← m_r · y_k;
9               else
                    /* f̄_k includes a_l in this case             */
10                  m_r ← m_r · ȳ_k;
11  end
```

The details of the procedure is as follows (let $f_3' = f$):

$$a_{001}, a_{000} : \Delta((001), (000)) = 3 \Rightarrow f'_3 \supset a_{001}$$

$$a_{001}, a_{011} : \Delta((001), (001)) = 2 \Rightarrow \bar{f}_2 \supset a_{001}$$

$$a_{001}, a_{101} : \Delta((001), (101)) = 1 \Rightarrow \bar{f}_1 \supset a_{001}$$

$$a_{001}, a_{110} : \Delta((001), (110)) = 1 \Rightarrow \bar{f}_1 \supset a_{001}$$

We have $m_1 = \bar{y}_1\bar{y}_2\bar{y}_3$.

$$a_{111}, a_{000} : \Delta((111), (000)) = 1 \Rightarrow f_1 \supset a_{111}$$

$$a_{111}, a_{011} : \Delta((111), (001)) = 1 \Rightarrow f_1 \supset a_{111}$$

$$a_{111}, a_{101} : \Delta((111), (101)) = 2 \Rightarrow f_2 \supset a_{111}$$

$$a_{111}, a_{110} : \Delta((111), (110)) = 3 \Rightarrow f_3 \supset a_{111}$$

So we have $m_2 = y_1y_2y_3$ and $f'_4 = m_1 + m_2$.

3.3.1.4 SPFD Global Rewiring (SPFD-GR)

The SPFD local rewiring scheme can only modify the connections to the fanin of the target node. Although it can be applied in LUT-based circuit optimization, it is rather limited to applications like LUT-based FPGA mapping and routing. SPFD global rewiring is a more generalized scheme using the concept of dominators, borrowed from the ATPG-based counterparts, and allows alternative wires to be added to more locations in the Boolean network while preserving the functions at the primary outputs.

SPFD-GR identifies possible alternative wires at the dominators. The general scheme is shown in Figure 3.8. Node N_D is a dominator of N_1, to which the target wire is connected, and the source node of the alternative wire is N_3 in the example. An important fact is that, though the scheme is named global, valid alternative wires are always added to the dominators of the target wire, where the logic change brought from target wire removal can be rectified by changing internal functions with wire addition.

In the SPFD-GR scheme, the target wire is temporarily removed from the network by assuming a constant logic 1 or 0 there. The functions of N_{t_2} and those nodes in its transitive fanouts are updated until the chosen dominator N_D is met. Basically, SPFD-GR will try the logic rectification at dominators one by one until all dominators are exhausted. If the function at N_D can be modified to satisfy the updated SPFD$[N_D]$, then the target wire can be removed directly from the network (tested by procedure *GR_Modify_Logic_1*). If not, a new wire w_a has to be formed by a node chosen from a candidate pool of candidate wires with N_D, such that a new internal function f_D can be set up to satisfy SPFD$[N_D]$. The internal function generation is carried in *GR_Modify_Logic_2*. The basic flow of the SPFD-GR algorithm is illustrated in Algorithm 3.11.

The core of the SPFD-GR algorithm is to determine whether it is possible to modify the internal logic of a node to satisfy its SPFD with or without an alternative wire. This is usually referred as the *node modification problem*. In the simpler case where an alternative wire is not needed, one verifies that all restricted MP-terms a_i are included by either function of SPFD$[(N_{t_2})]$. Then we can construct a proper internal function to distinguish

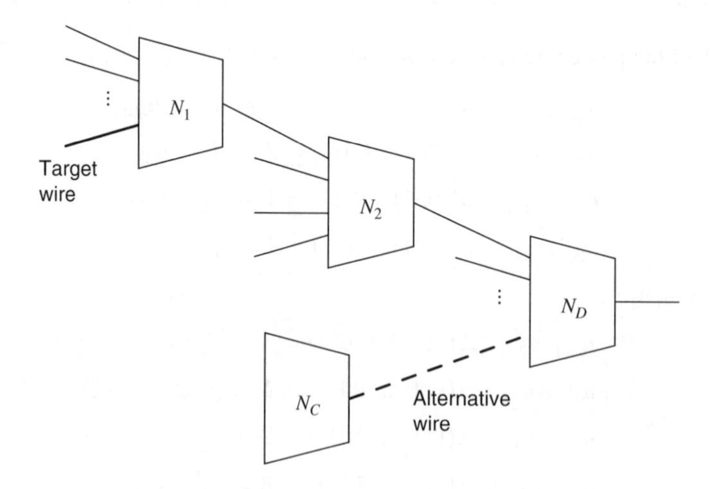

Figure 3.8 SPFD-based global rewiring

Algorithm 3.11: Procedure *GR_Main*

```
   input  : target wire w_t = (N_{t_1}, N_{t_2}), dominator G_D
   output: result: success or fail, alternative wire w_a
1  begin
2    Temporarily remove w_t;
3    Update the output function f(N_{t_2});
4    Propagate new output function to transitive fanouts till N_D;
5    if GR_Modify_Logic_1(N_{t_2}) = success then
6      f_{t_2} <- f'_i;
7      update SPFDs;
8      return success;
9    foreach candidate source N_C do
10     if GR_Modify_Logic_2(N_{t_2}, N_C) = success then
11       w_a <- (N_C, N_D), commit w_a;
12       f_{t_2} <- f'_i;
13       update SPFDs;
14       return success;
15   Restore all function modifications;
16   return fail;
17 end
```

the SPFD. Such construction is similar to that in *LR_Modify_Logic*, and is outlined in *GR_Modify_Logic_1* (Algorithm 3.12). The set $\mathbf{B_0}$ and $\mathbf{B_1}$ contain the nonempty MP-terms that are included by f_{ON} and f_{OFF}, respectively. Those MP-terms that are not included will be saved in the set $\mathbf{B_2}$. When it is empty, then we can generate a new internal function as the Boolean sum of the terms in $\mathbf{B_1}$ to distinguish SPFD$[N_{t_2}]$.

If $\mathbf{B_2}$ is not empty, then an alternative wire is required. In Algorithm 3.13, the constructions of sets $\mathbf{B_0}$, $\mathbf{B_1}$, and $\mathbf{B_2}$ are similar. We are interested to see whether the new fanin N_C can help distinguish the SPFD when it is connected to the dominator N_D. MP-terms in $\mathbf{B_1}$ and $\mathbf{B_0}$ are already included without the new fanin, so we can focus on the terms of $\mathbf{B_2}$ and see whether they can all be included by g_{n+1}, the function of the new fanin. If not, then (N_C, N_D) cannot give a valid alternative wire; otherwise we can confirm an alternative wire and generate a new internal function for N_D.

Let us consider an example of *GR_Modify_Logic_2* in Figure 3.9: a dominator N_D has SPFD $(f_{ON} = x_1 + x_2, f_{OFF} = \bar{x}_1\bar{x}_2)$. Originally it has two fanins N_1 and N_2, and the functions there are $f_1 = \bar{x}_1 x_2$ and $f_2 = x_1\bar{x}_2$, respectively. Assume that the effect of target wire removal has been propagated properly till N_D.

We first compute the sets $\mathbf{B_0}$, $\mathbf{B_1}$, and $\mathbf{B_2}$, as follows:

$$a_0 = \bar{f}_1\bar{f}_2(f_{ON} + f_{OFF}) = \bar{x}_1\bar{x}_2 + x_1x_2$$

$$a_1 = \bar{f}_1 f_2(f_{ON} + f_{OFF}) = x_1\bar{x}_2 \leq f_{ON}$$

$$a_2 = f_1\bar{f}_2(f_{ON} + f_{OFF}) = \bar{x}_1 x_2 \leq f_{ON}$$

Algorithm 3.12: Procedure *GR_Modify_Logic_1*

```
input : node N_i with SPFD[(N_i)] = (f_ON, f_OFF)
output: result: success or fail, f'_i: new internal functions for N_i
1  begin
       /* 2^p local function products / MP-terms                */
2      B ← {b_1, b_2, ..., b_{2^p}};
3      A ← {a_i = b_i(f_ON + f_OFF)};
4      B_1 ← {b_i | nonempty a_i ≤ f_ON};
5      B_0 ← {b_i | nonempty a_i ≤ f_OFF};
6      B_2 ← B − B_1 − B_0;
7      if B_2 = φ then
8          f'_i = Σ_{b_i ∈ B_1} b_i;
9          return success;
10     return fail;
11 end
```

Algorithm 3.13: Procedure *GR_Modify_Logic_2*

```
input : node N_i with SPFD[(N_i)] = (f_ON, f_OFF), candidate node N_C
output: result: success or fail, f'_i: new internal functions for N_i
1  begin
       /* 2^p local function products / MP-terms                */
2      B ← {b_1, b_2, ..., b_{2^p}};
3      A ← {a_i = b_i(f_ON + f_OFF)};
4      B_1 ← {b_i | nonempty a_i ≤ f_ON};
5      B_0 ← {b_i | nonempty a_i ≤ f_OFF};
6      B_2 ← B − B_1 − B_0;
7      p, q ← 0;
8      foreach β ∈ B_2 do
9          r_0 = ¬g_{n+1} β(f_ON + f_OFF);
10         r_1 = g_{n+1} β(f_ON + f_OFF);
11         if r_1 ≤ f_ON and r_0 ≤ f_OFF then
12             p ← p + β g_{n+1};
13         else if r_0 ≤ f_ON and r_1 ≤ f_OFF then
14             q ← q + β ¬g_{n+1};
15         else
16             return fail;
       /* a valid alternative wire can be formed with the
          candidate node N_C                                      */
17     f'_i = Σ_{b_i ∈ B_1} b_i + p + q;
18     return success;
19 end
```

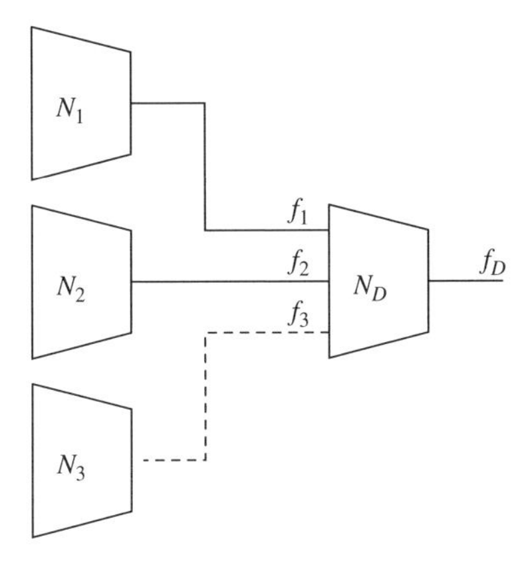

Figure 3.9 SPFD-GR: *GR_Modify_Logic_2* Example

$$a_3 = f_1 f_2 (f_{ON} + f_{OFF}) = 0$$
$$\mathbf{B_0} = \phi$$
$$\mathbf{B_1} = \{b_1, b_2\}$$
$$\mathbf{B_2} = b_0$$

As $\mathbf{B_2}$ is not empty, an alternative wire is needed. Suppose we choose a candidate node N_3 with function $f_3 = x_1 x_2$. A check of f_3 with $\mathbf{B_2}$ is

$$\bar{f}_1 \bar{f}_2 \bar{f}_3 = \bar{x_1} \bar{x_2} \leq f_{OFF}, \bar{f}_1 \bar{f}_2 f_3 = x_1 x_2 \leq f_{ON}$$

Therefore, (N_3, N_D) is a valid alternative wire, and the required new function of N_D should be

$$f'_D = \bar{f}_1 f_2 + f_1 \bar{f}_2 + \bar{f}_1 \bar{f}_2 f_3$$

3.3.1.5 Enhanced SPFD Global Rewiring (SPFD-ER)

Recall that, in the SPFD propagation scheme, when a node has multiple fanouts, we are going to take the union of the functions from the fanouts' SPFDs (step 9 in Algorithm 3.6). This simplifies the node SPFD to a single pair for efficient atomic assignment to its fanin pins. However, the union operation creates unnecessary constraints between disjunctive SPFD pairs and reduce the chance of finding an alternative wire. The following example explains the concept.

Node G in Figure 3.10 has two fanouts whose SPFDs are $\{(ab, \bar{a}b)\}$ and $\{(a\bar{b}, \bar{a}\bar{b})\}$, respectively. In the original procedure, we are going to union the function pairs

$$\text{SPFD}[G] = \{(ab + a\bar{b}, \bar{a}b + \bar{a}\bar{b})\} = \{(a, \bar{a})\}$$

As a result, only functions $f = a$ or $f = \bar{a}$ can satisfy SPFD$[G]$. But this is not necessary since functions like $f = a \oplus b$ or $f = a \bar{\oplus} b$ can also satisfy both wire SPFDs at the fanouts. The first enhancement included in the SPFD-ER scheme is an algorithm to determine whether there exists a function that distinguishes all pairs in an SPFD.

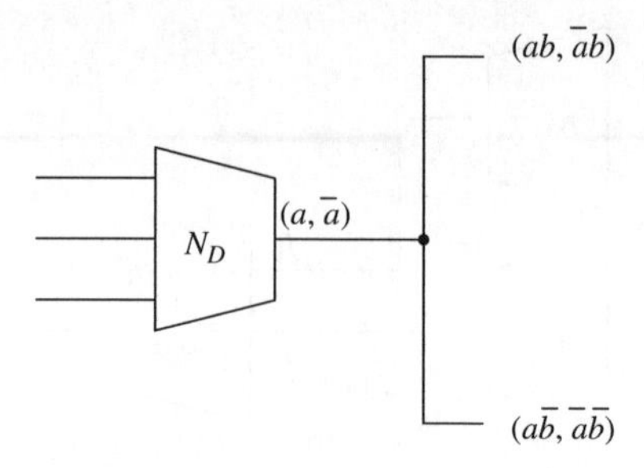

Figure 3.10 SPFD-ER: Example

Given a node N, we collect all function pairs to be distinguished from its fanouts. Let the pairs be $(g_{1a}, g_{1b}), (g_{2a}, g_{2b}), \ldots, (g_{ka}, g_{kb})$, and let G denote the care-set sum of all functions:

$$G = \sum_{i=1}^{k} (g_{ia} + g_{ib})$$

. Then we construct the MP-terms using p fanin functions of N (denoted by $g_1, g_2, \ldots$) and restrict them with G:

$$\beta_0 = b_0 \cdot G = \bar{g}_1 \bar{g}_2 \ldots \bar{g}_n \cdot G$$

$$\beta_1 = b_1 \cdot G = \bar{g}_1 \bar{g}_2 \ldots g_n \cdot G$$

$$\vdots$$

$$\beta_{2^p} = b_{2^p} \cdot G = g_1 g_2 \ldots g_n \cdot G$$

.

We build a graph S with 2^p nodes: each vertex represents each β and there is an edge (a, b) between vertices a and b if, for some i, $\beta_a \cap g_{ia} \neq \phi$ and $\beta_b \cap g_{ib} \neq \phi$.

Theorem 3.9 *S is bipartite if and only if there exists a function $f(g_1, g_2, \ldots, g_n)$, which can distinguish $(g_{1a}, g_{1b}), (g_{2a}, g_{2b}), \ldots, (g_{ka}, g_{kb})$.*

Checking whether a graph is bipartite can be done in linear time using depth-first search, and this imposes little penalty in the efficiency of the rewiring algorithm. It should be noted the checking for bipartiteness is used for alternative wire identification but not SPFD propagation, since the number of function pairs is too large without union of the functions. The updated algorithm flow is shown in Algorithm 3.14.

The second improvement in the SPFD-ER scheme is the introduction of several useful heuristics for assignments of atomic SPFD pairs. Recall that in Algorithm 3.9) we decompose fanout SPFDs into an atomic SPFD and assign it to an appropriate fanin, as long as the fanin can satisfy the pair. It is natural that when there is fewer function pairs in an SPFD, it is easier to be satisfied. In other words, the wire has a greater chance to be replaced as there are fewer constraints.

Algorithm 3.14: Procedure *ER_Main*

```
    input : target wire w_t = (N_{t_1}, N_{t_2}), dominator G_D
    output: result: success or fail, alternative wire w_a
 1  begin
 2      Temporarily remove w_t;
 3      Update the output function f(N_{t_2});
 4      Propagate new output function to transitive fanouts till N_D;
 5      Build graph S_D;
 6      if S_D is bipartite then
 7          f_{t_2} ← f'_i;
 8          Update SPFDs;
 9          return success;
10      foreach candidate source N_C do
11          Add wire (N_C, G_D) to the network;
12          Build graph S_D;
13          if S_D is bipartite then
14              w_a ← (N_C, N_D), commit w_a;
15              f_{t_2} ← f'_i;
16              Update SPFDs;
17              return success;
18      Restore all function modifications;
19      return fail;
20  end
```

Delay and area are the common objectives in technology-independent optimization. For delay optimization, SPFD-ER aims at reducing the longest delay path. Therefore the fanins are sorted against the timing criticality, and the more critial fanin will be assigned fewer function pairs. For area optimization, fewer function pairs will be assigned to fanin wires having less number of fanouts, making it more likely to be removed by alternative wires.

Another useful heuristic in the atomic SPFD assignment is to consider the relative size of the SPFDs. Suppose two fanins N_p and N_q whose functions can both distinguish an atomic pair (f_a, f_b). If SPFD$[N_p]$ includes f_a and SPFD$[N_q]$ does not include any of f_a or f_b, then the pair will be assigned to N_p because the size of SPFD$[N_p]$ will remain unchanged, and the size of SPFD$[N_q]$ increases otherwise.

3.3.1.6 Quantitative Comparisons of SPFD-Based Rewiring Algorithms

Rewiring ability is the basic metric in comparing rewiring algorithms. It is defined to be the number of wires having at least one alternative wire. Table 3.1 shows the results of an experiment comparing the three main SPFD-based rewiring algorithms (i.e., SPFD-LR, SPFD-GR, and SPFD-ER). It can be seen that SPFD-ER has 70% more target wires with alternative wires when compared to SPFD-LR and 18% more when compared directly to SPFD-GR.

Table 3.1 Comparison of rewiring ability of SPFD-based rewiring algorithms for four-LUT FPGA designs

	Total no. of wires	SPFD-LR	SPFD-GR	GR over LR (%)	SPFD-ER	ER over LR (%)
C1908	423	74	99	33.8	114	54.1
C432	538	154	183	18.8	208	35.1
alu4	939	277	419	51.3	522	88.4
apex6	1025	270	345	27.8	410	51.9
dalu	1338	468	704	50.4	739	57.9
example2	433	85	136	60.0	169	98.8
term1	244	71	99	39.4	114	60.6
x1	557	164	222	35.4	271	65.2
x3	958	154	319	107.1	366	137.7
alu2	510	179	253	41.3	306	70.9
C5315	1772	500	607	21.4	757	51.4
Average				44.3		70.2

Table 3.2 Comparison of rewiring ability for different atomic SPFD assignment methods

Circuits	A	B	C	D
C1908	31	34	35	35
C432	26	22	26	26
alu4	50	43	55	56
apex6	34	32	35	34
dalu	59	50	57	57
example2	11	14	13	13
term1	4	4	4	4
x1	14	17	18	18
x3	21	25	25	25
alu2	48	50	56	53
Average (%)		3.1	10.5	9.8

Method A: Randomly assigns to a fanin that can distinguish the pair.
Method B: Randomly sorts the fanins and assign the first one that can distinguish the pair.
Method C: Delay-oriented heuristic, sorts against criticality.
Method D: Method C plus SPFD-size-oriented heuristic.

The rewiring ability of SPFD-based rewiring algorithms is affected by the choice of atomic SPFD pair assignment, as explained in the previous subsection. The experimental results in Table 3.2 show the comparisons of the commonly used assignment heuristic. It is clear that criticality-oriented heuristics perform much better than random ones. It is reported that SPFD-ER with method C can have 38.1% more rewiring power than SPFD-GR with method A.

3.3.1.7 Partitioning for Large Circuits

BDDs are a common means to represent logic functions, so SPFDs are also represented by BDDs. The runtime of SPFD-based rewiring algorithms is dominated by the efficiencies of the BDD operations. In large circuits, it is usually impossible to build BDDs for all internal node functions using global (input) variables. Partitioning has to be used with the rewiring algorithm to preserve time efficiency in alternative wire identification. Luckily, this would not impact the overall rewiring ability, as it is reported that most alternative wires are not far away from the target wires. In experiments, SPFD-GR can run 2.6× faster with only a 2.3% decrease in the rewiring ability. It is recommended to include tools for partitioning when SPFD-based rewiring algorithms are used on large-scale or industrial designs.

References

S. C. Chang and M. Marek-Sadowska. Perturb and simplify: multilevel Boolean network optimizer. *IEEE Transactions on Computer-Aided Design of Integrated Circuits and Systems*, 15(12):1494–1504, 1996.

C.-W. J. Chang and M. Marek-Sadowska. Who are the alternative wires in your neighborhood? (alternative wires identification without search). In *Proceedings of the 11th Great Lakes symposium on VLSI*, GLSVLSI '01, pages 103–108, New York, 2001. ACM. ISBN: 1-58113-351-0. doi: 10.1145/368122.368880.

S.-C. Chang, L. P. P. P. van Ginneken, and M. Marek-Sadowska. Fast Boolean optimization by rewiring. In *Computer-Aided Design, 1996. ICCAD-96. Digest of Technical Papers., 1996 IEEE/ACM International Conference on*, pages 262–269, November 1996. doi: 10.1109/ICCAD.1996.569641.

S.-C. Chang, L. P. P. P. van Ginneken, and M. Marek-Sadowska. Circuit optimization by rewiring. *IEEE Transactions on Computers*, 48(9):962–970, 1999. ISSN: 0018-9340. doi: 10.1109/12.795224.

Y.-C. Chen and C.-Y. Wang. Fast detection of node mergers using logic implications. In *Computer-Aided Design - Digest of Technical Papers, 2009. ICCAD 2009. IEEE/ACM International Conference on*, pages 785–788, November 2009.

Y.-C. Chen and C.-Y. Wang. Fast node merging with don't cares using logic implications. *IEEE Transactions on Computer-Aided Design of Integrated Circuits and Systems*, 29(11), 2010a.

Y.-C. Chen and C.-Y. Wang. Node addition and removal in the presence of don't cares. In *Design Automation Conference (DAC), 2010 47th ACM/IEEE*, pages 505–510, 2010b.

Y.-C. Chen and C.-Y. Wang. Logic restructuring using node addition and removal. *IEEE Transactions on Computer-Aided Design of Integrated Circuits and Systems*, 31(2), 2012.

K.-T. Cheng and L. A. Entrena. Multi-level logic optimization by redundancy addition and removal. In *Design Automation, 1993, with the European Event in ASIC Design. Proceedings. [4th] European Conference on*, pages 373–377, February 1993. doi: 10.1109/EDAC.1993.386447.

L. A. Entrena and K.-T. Cheng. Combinational and sequential logic optimization by redundancy addition and removal. *IEEE Transactions on Computer-Aided Design of Integrated Circuits and Systems*, 14(7):909–916, 1995. ISSN: 0278-0070. doi: 10.1109/43.391740.

M. A. Iyer and M. Abramovici. FIRE: a fault-independent combinational redundancy identification algorithm. *IEEE Transactions on Very Large Scale Integration (VLSI) Systems*, 4(2):295–301, 1996. ISSN: 1063-8210. doi: 10.1109/92.502203.

S. M. Plaza, K.-H. Chang, I. L. Markov, and V. Bertacco. Node mergers in the presence of don't cares. In *ASP-DAC '07: Proceedings of the 2007 Asia and South Pacific Design Automation Conference*, pages 414–419, Washington, DC, 2007. IEEE Computer Society. ISBN: 1-4244-0629-3. doi: 10.1109/ASPDAC.2007.358021.

W. C. Tang, W. H. Lo, T. K. Lam, K. K. Mok, C. K. Ho, S. H. Yeung, H. B. Fan, and Y. L. Wu. A quantitative comparison and analysis on rewiring techniques. In *ASIC, 2003. Proceedings. 5th International Conference on*, Volume 1, pages 242–245, 2003. doi: 10.1109/ICASIC.2003.1277533.

Q. Zhu, N. Kitchen, A. Kuehlmann, and A. Sangiovanni-Vincentelli. SAT sweeping with local observability don't-cares. In *DAC '06: Proceedings of the 43rd Annual Design Automation Conference*, pages 229–234, New York, 2006. ACM. ISBN: 1-59593-381-6. doi: 10.1145/1146909.1146970.

4

Delete-First Rewiring Techniques

Redundancy addition and removal (Cheng and Entrena 1993, Entrena and Cheng 1995, Chang and Marek-Sadowska 1996, 2001, Chang et al. 1999) is an automatic test pattern generation (ATPG)-based restructuring technique that works by adding new redundant wires to a circuit to make the target wires in the circuit redundant and removable. Given an irredundant target wire w_t in a circuit C, w_a is an alternative wire for w_t if both w_t and w_a are redundant in circuit $C + w_a$. We use the symbol $+$ to represent addition and the symbol $-$ to represent removal. This relation can be expressed symbolically as follows:

$$C \equiv C + w_a \equiv C + w_a - w_t \tag{4.1}$$

Is it necessary that the pairs of target and alternative wires are mutually redundant? In other words, is it possible to have the following relation?

$$C \equiv C - w_t + w_a \tag{4.2}$$

Recent research has proven that this is possible. It has been found that an alternative wire does not have to be redundant and does not have to make the target wire redundant as long as the original circuit function is maintained after the removal of the target wire and the addition of the alternative wire. Since some errors are introduced during the transformation process but they do not have any effect on the circuit function in the end, this process is known as *error cancellation*. This kind of rewiring techniques is known as *delete-first rewiring* because target wires can be removed first and the alternative wires can be identified and the circuit functions can be rectified later. The concept of error cancellation can be represented graphically as shown in Figure 4.1. In the figure, errors $e1$ and $e2$ cancel each other before reaching the primary outputs (POs). The removal of a target wire and the addition of an alternative wire can be thought of as errors $e1$ and $e2$.

The oldest developed concept that is similar to error cancellation was introduced in Muroga et al. (1989). In their work, the authors suggested a circuit transformation approach known as *error compensation* that takes the removal of gates into account. They adopted the concepts that are similar to those applied in SPFD-based rewiring techniques (Section 2.3.2).

Boolean Circuit Rewiring: Bridging Logical and Physical Designs, First Edition.
Tak-Kei Lam, Wai-Chung Tang, Xing Wei, Yi Diao and David Yu-Liang Wu.

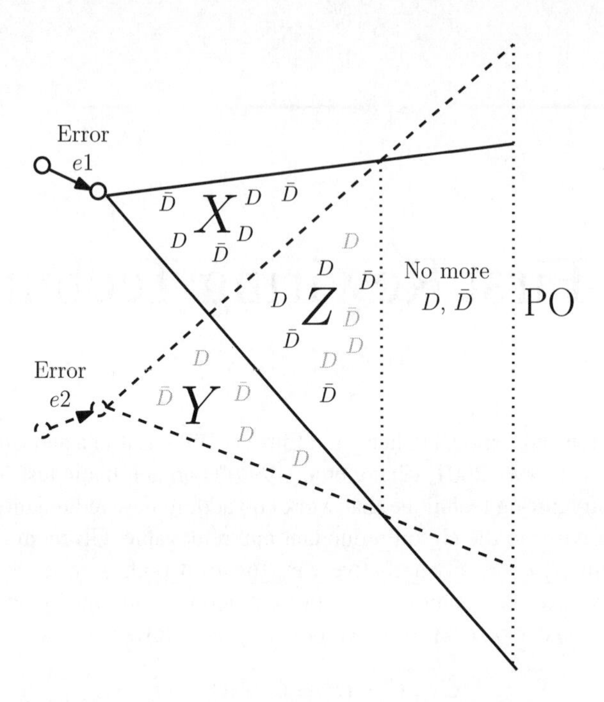

Figure 4.1 General abstract view of error cancellation

In Chang and Marek-Sadowska (2007), the authors proposed the first formal symbolic model that describes the concept of error cancellation. We paraphrase the statement of the error cancellation model as Theorem 4.1.

Theorem 4.1 *The removal of an irredundant wire w_t in a circuit introduces an error to the circuit. The addition of an irredundant wire w_a introduces another error to the circuit. These two errors are denoted as $\epsilon_r(w_t)$ and $\epsilon_a(w_a)$, respectively. They are observable if their effects can be propagated to any one of the primary outputs. Obviously, the function of the circuit with error $\epsilon_r(w_t)$ only is different from the original function and the error $\epsilon_r(w_t)$ is observable. The circuit with error $\epsilon_a(w_t)$ only also differs from the original circuit. If the circuit having error $\epsilon_r(w_t)$ is added with an irredundant wire w_a, it has both errors $\epsilon_r(w_t)$ and $\epsilon_a(w_a)$. Wire w_a is an alternative wire for wire w_t if and only if these two errors cancel each other completely before being propagated to any primary outputs for any test vectors.*

The general theoretical aspects of error cancellation were addressed Chang and Marek-Sadowska (2007). However, the authors made no experimental justification for their approach. From theoretical analysis, it can be understood that their approach is quite intensive computationally and is less practical for large circuits.

4.1 IRRA

A practical implementation of error-cancellation-based rewiring was proposed in Lin and Wang (2009). It is known as irredundancy removal and addition (IRRA) and was introduced in (Section 2.2.2.1) of this book. In this approach, the error that has been caused by removal of a target wire in a circuit is first corrected by adding additional logic (alternative logic) into the circuit. The alternative logic is then reduced to a wire so that the end result is the substitution of a wire by a wire.

With regard to a stuck-at fault, the authors of IRRA classify a subset out of the whole set of mandatory assignments (MAs), which is useful for formulating their theories and approach of rewiring.

Definition 4.1 *A source mandatory assignment (SMA) for a stuck-at fault in a circuit is a mandatory assignment whose transitive fanin cone contains no other mandatory assignments.*

This definition is based on the structure of the circuit. In Figure 4.2, the mandatory assignments for the fault $sa0(f \rightarrow g)$ are shown. Gate g is a dominator of the fault, and its side input is assigned with its noncontrolling value 0. Then, it can be implied that both a and b are 0. The fault requires setting wire $f \rightarrow g$ to 1. Therefore, c and d have to be assigned with 1. The value assignments $\{e = 0, f = 1\}$ are not SMAs because their transitive fanin cone contains other mandatory assignments. In this example, the SMAs are $\{a = 0, b = 0, c = 1, d = 1\}$.

The definition of SMAs is used in the construction of exact addition network (EAN) and exact removal network (ERN), which are described as follows.

Definition 4.2 *Given a target wire and a destination gate d in a circuit. The EAN at gate d is a Boolean network composed of minterms whose values will change from 0 to 1 after the removal of the target wire.*

Definition 4.3 *Given a target wire and a destination gate d in a circuit. The ERN at gate d is a Boolean network composed of minterms whose values will change from 1 to 0 after the removal of the target wire.*

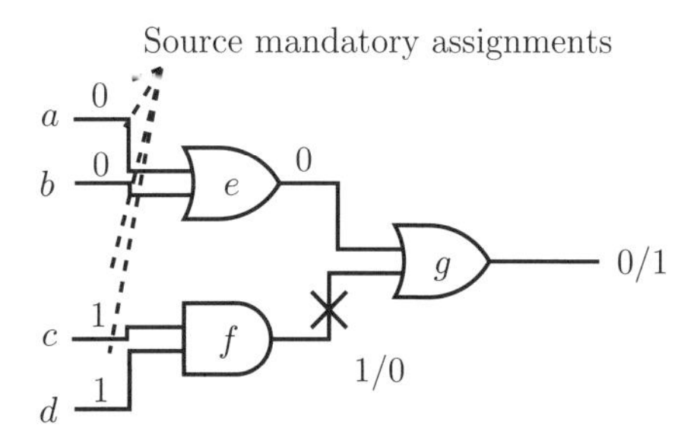

Figure 4.2 Source mandatory assignments

These two special Boolean networks can be constructed according to the following theorems:

Theorem 4.2 *Suppose a target wire w_t is to be removed from a circuit C_g. After the removal of the target wire, the new circuit is C_f. The SMAs of the corresponding stuck-at fault at the target wire is SMA_{w_t}. Then, the cofactors of the function implemented at a destination gate d in C_g and C_f with respect to SMA_{w_t} are $d_{C_g}|_{SMA_{w_t}}$ and $d_{C_f}|_{SMA_{w_t}}$, respectively. Let $AND(SMA_{w_t})$ be the logical conjunction of all the SMAs.*
The EAN at gate d is

$$EAN = AND(SMA_{w_t}) \cdot \overline{d_{C_g}|_{SMA_{w_t}}} \cdot d_{C_f}|_{SMA_{w_t}} \tag{4.3}$$

The ERN at gate d is

$$ERN = AND(SMA_{w_t}) \cdot d_{C_g}|_{SMA_{w_t}} \cdot \overline{d_{C_f}|_{SMA_{w_t}}} \tag{4.4}$$

Then, the error caused by the removal of the target wire w_t can be corrected using the EAN and ERN according to Theorem 4.3. These two Boolean networks form a rectification network. This theorem can be visually presented as in Figure 4.3.

Theorem 4.3 *After the removal of the target wire w_t, the original function implemented by gate d can be implemented by $(d + ERN) \cdot EAN$, where EAN and ERN are, respectively, the exact addition network and exact removal network with respect to wire w_t at gate d.*

The authors proved that the Boolean network $d_{C_f}|_{SMA_{w_t}}$ is in fact redundant and can be removed. Figure 4.4 illustrates the simplified rectification network.

Theorem 4.4 *Referring to Theorem 4.2, the EAN at gate d can be simplified to*

$$EAN = AND(SMA_{w_t}) \cdot \overline{d_{C_g}|_{SMA_{w_t}}} \tag{4.5}$$

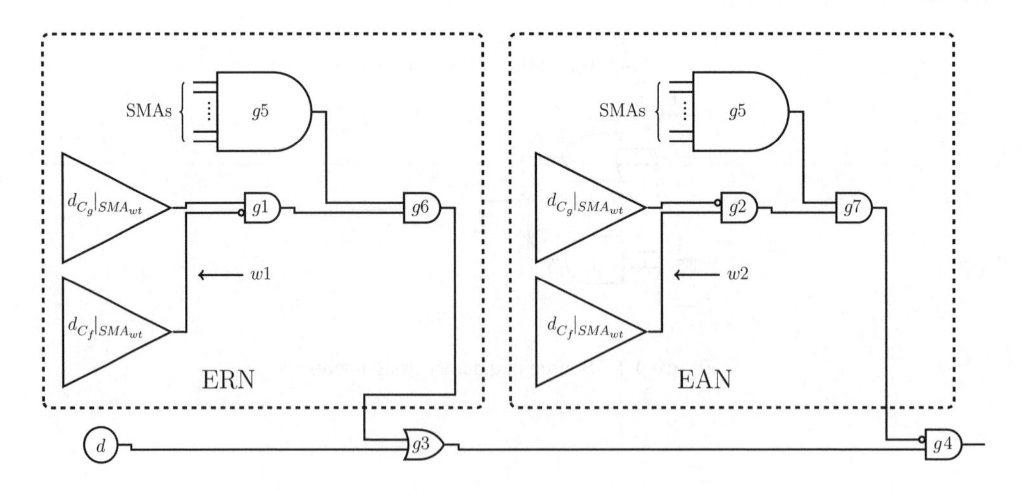

Figure 4.3 Rectification network

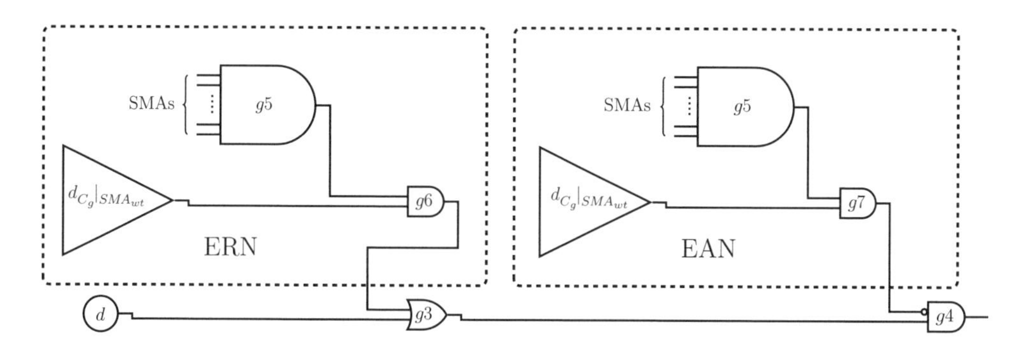

Figure 4.4 Simplified rectification network

The ERN at gate d can be simplified to

$$ERN = AND(SMA_{w_t}) \cdot d_{C_g}|_{SMA_{w_t}} \tag{4.6}$$

Proof. The proof is equivalent to proving that the wires $w1$ and $w2$ in Figure 4.3 are redundant. With regard to the test for fault $sa0(w1)$, the mandatory assignments are $\{d_{C_g}|_{SMA_{w_t}} = 1, d_{C_f}|_{SMA_{w_t}} = 1, g5 = 1, d = 0, g7 = 0\}$. Since both $d_{C_g}|_{SMA_{w_t}} = 1$ and $d_{C_f}|_{SMA_{w_t}} = 1$, the value of the gate d is 1. Then, there is a contradiction. Therefore, wire $w1$ is redundant.

Similarly, there are conflicts in the mandatory assignments for fault $sa1(w2)$. Wire $w2$ is also redundant, and the rectification network can be simplified by removing wires $w1$ and $w2$. ■

As readers may have noticed, the purpose of constructing the EAN and ERN with respect to a target wire is to create alternative wires for the target wire. According to Theorem 4.3, the function of gate d after removing the target wire can be restored using $(d + ERN) \cdot EAN$. We have already derived the right tool or mechanism to rectify errors. The next question is: where are the errors?

4.1.1 Destination of Alternative Wires

Nodes may receive values $1/0(D)$ or $0/1(\bar{D})$ during the activation and propagation of a fault. Obviously, these nodes have errors after the removal of the target wire. This suggests that a rectification network has to be constructed for every node with mandatory assignments D or $\bar{D}$. In the IRRA approach, these nodes are considered one of the two possible destinations of alternative wires.

Referring to Figure 4.4, network $d_{C_g}|_{SMA_{w_t}}$ is 1 and the EAN is evaluated to be constant 0 if its corresponding gate has the mandatory assignment D. For a gate with mandatory assignment $\bar{D}$, network $d_{C_g}|_{SMA_{w_t}}$ is 0, and the ERN is evaluated to a constant 0. Hence, its rectification network can be further simplified. The resultant rectification networks for the destination gate with mandatory assignments D and $\bar{D}$ are shown in Figure 4.5.

Besides choosing gates with mandatory assignments D or $\bar{D}$ as the destination nodes of alternative wires, gates with forced mandatory assignment (FMA)s 0 or 1 are also chosen.

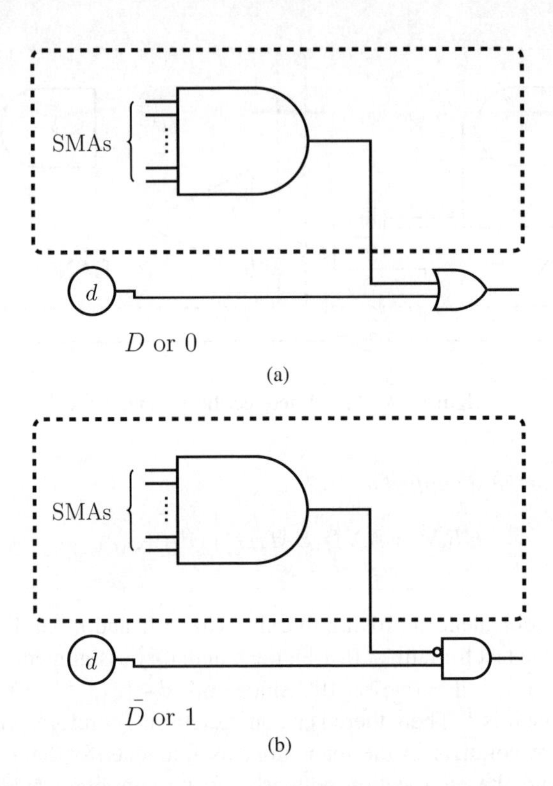

Figure 4.5 Destination nodes of alternative wires. (a) Rectification network for destination gate with MA value D or 0; and (b) $\bar{D}$ or 1

The rationale is the same as that for REWIRE, RAMFIRE, and other add-first or RAR-based rewiring techniques. That is, we want to change the value of a FMA to its opposite such that the fault caused by the removal of the target wire becomes untestable. In this case, the correctness of the alternative wires found have to be verified (this is not mentioned in (Lin and Wang 2009)). Verification of an alternative wire can be done by checking whether there are any error effects observable at any primary outputs after removing the target wire and adding the alternative wire.

In IRRA, a rectification network constructed from SMAs is required no matter whether dominators or gates with forced mandatory assignments are used as the destination nodes of alternative wires. Since it is the logical conjunction of all SMAs, its size may be very large. Clearly, using the rectification networks as the sources of alternative wires is not desirable. The authors of IRRA have found a way to trim the rectification networks.

4.1.2 Source of Alternative Wires

After studying the properties of SMAs associated with a target wire carefully, it is found that they can be further classified into three categories. A SMA can be redundant or irredundant. The authors of IRRA also define a third kind of SMA as semi-redundant SMA. They are explained it as follows:

Redundant source mandatory assignments

They are mandatory assignments that are redundant after the removal of the target wire w_t. This means their corresponding inputs to $AND(SMA_{w_t})$ can be removed.

Irredundant source mandatory assignments

They are mandatory assignments that are not redundant after the removal of the target wire w_t. Their corresponding inputs to $AND(SMA_{w_t})$ can also be removed.

Semiredundant source mandatory assignments

Similar to redundant SMAs, they are mandatory assignments that are redundant after the removal of the target wire w_t. The difference between the two is that the semi-redundant SMA is redundant only when the values of some of the irredundant SMAs are set to some constants.

Let us use the example designed by the authors of IRRA to illustrate the above concepts. An example circuit is shown in Figure 4.6. In the circuit, wire $g1 \rightarrow g2$ is the target wire. Suppose that the target wire and the associated gates and wires (drawn in dashed lines) have been removed. As we can see, gate $g5$ is a dominator of fault $sa1(g1 \rightarrow g2)$ and its value assignment is $\bar{D}$. The SMAs are found to be $\{a = 0, b = 0, c = 1, e = 1\}$. Based on the analysis in Section 4.1.1, a rectification network can be constructed as shown in the figure. In order to simplify the rectification work, the redundancy of each of the SMAs has to be evaluated. For $a = 0$, we have to test fault $sa0(a \rightarrow g10)$ because the corresponding input to gate $g10$ is inverted. It can be easily figured out that fault $sa0(a \rightarrow g10)$ is testable. Hence, $a = 0$ is an irredundant SMA. Likewise, $b = 0$ is also irredundant. With regard to $c = 1$, the corresponding fault $sa1(c \rightarrow g10)$ is found to be redundant. There is a conflict, $a = 0$ and $a = 1$, in the test. Since $a = 0$ is an irredundant SMA, $c = 1$ is regarded as semi-redundant. In the test for

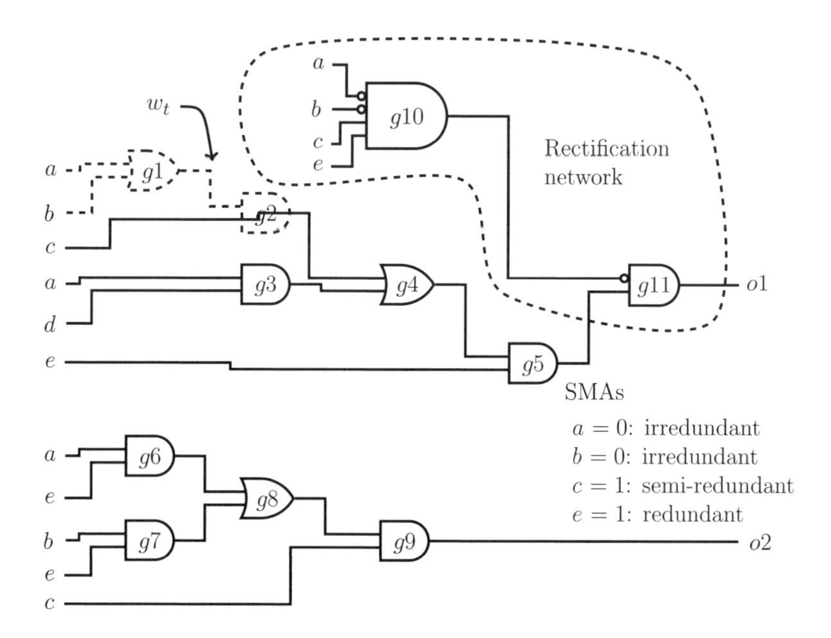

Figure 4.6 Source mandatory assignments of fault $sa1(g1 \rightarrow g2)$

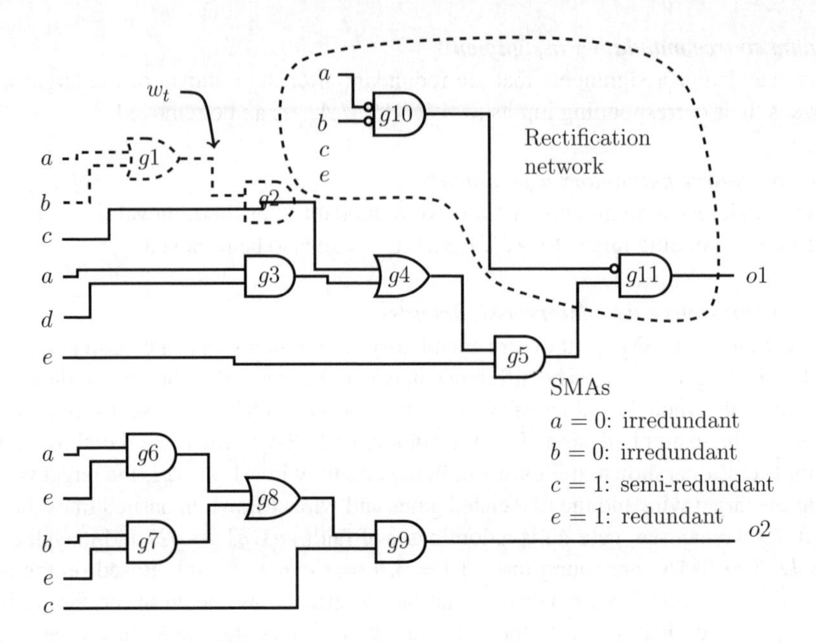

Figure 4.7 Removal of semi-redundant and redundant SMAs

fault $sa1(e \to g10)$, e is set to 0 to activate the fault. Then, the values of both gates $g5$ and $g11$ are 0, which means $sa1(e \to g10)$ can never be propagated to the primary outputs. Hence, the only redundant SMA is $e = 1$.

This categorization of SMAs allows us to simplify the rectification network. Semi-redundant and redundant SMAs can be removed from $AND(SMA_{w_t})$. The circuit after removing (semi-)redundant SMAs is shown in Figure 4.7. Irredundant SMAs may also be substituted or removed according to the rules illustrated in Figure 4.8.

Suppose the input a in the figure is an irredundant SMA that we would like to remove. Furthermore, suppose all other inputs to the gates are mandatory assignments unless otherwise specified. In Figure 4.8(a), it is obvious that the SMA $a = 1$ can be backwardly implied from the mandatory assignment $g1 = 1$. Then, $g1 = 1$ is said to be a substitute for $a = 1$. In Figure 4.8(b), the mandatory assignment $g1 = 0$ cannot substitute $a = 0$ because c is not a mandatory assignment and its value is undetermined. For the case of Figure 4.8(c), the backward implication is simply invalid. The backward implication from a gate's controlling value at its output to one of the gate's input is valid only if all other inputs of the gate have non-controlling and redundant SMAs, as shown in Figure 4.8(d). The redundancy requirement is necessary because the semiredundant SMAs may depend on the target irredundant SMA. For example, in Figure 4.8(e), the semiredundant SMA f may be dependent on the value of a. Thus, the irredundant SMA a has to be maintained.

Irredundant SMAs may share the same set of substitutes. Referring to the example in Figure 4.7, $g6 = 0$ is a substitute for $a = 0$ because $e = 1$ is a redundant SMA. The set of substitutes for $a = 0$ can be found to be $\{g6 = 0, g8 = 0\}$. Then, the set of substitutes

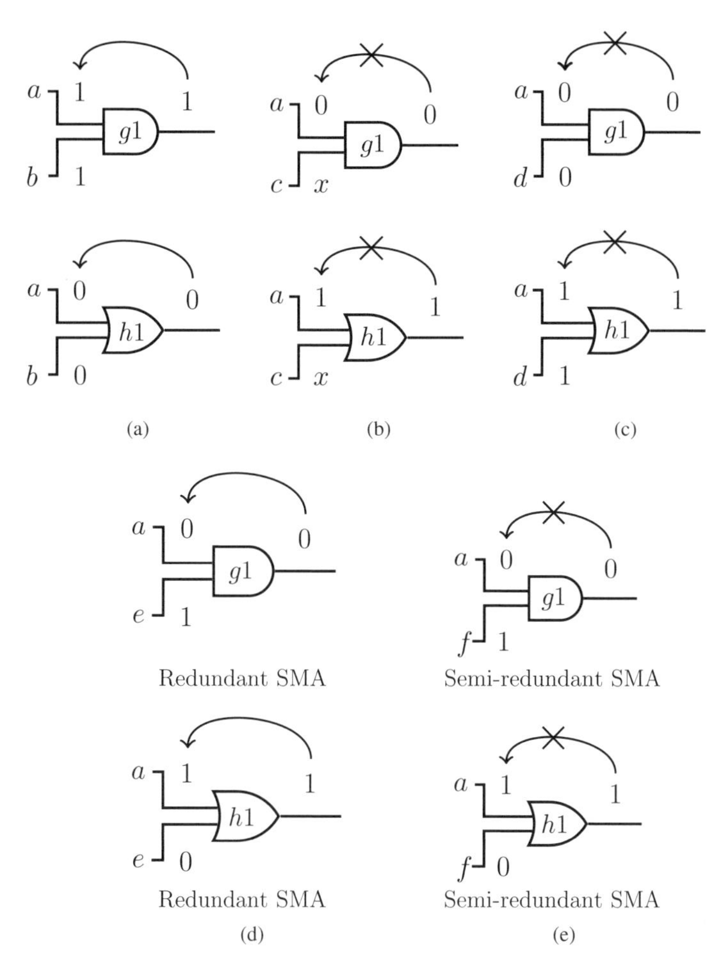

Figure 4.8 Substitution rules of source mandatory assignments for AND and OR gates. (a) Rule 1; (b) Rule 2; (c) Rule 3; (d) Rule 4; (e) Rule 5;

for $b = 0$ can be found to be $\{g7 = 0, g8 = 0\}$. Hence, the common set of substitutes for both $a = 0$ and $b = 0$ is $\{g8 = 0\}$. From this analysis, we have found out that $g8 = 0$ can substitute the rectification network $g10 = \bar{a} = \bar{b}$. In other words, $g8 \rightarrow g11$ can be used to replace the rectification work, and is an alternative wire for the target wire $g1 \rightarrow g2$. The rewired circuit is depicted in Figure 4.9.

We have explained the method of identifying the sources of alternative wires through an example. First, a rectification network is constructed at the dominators or the nodes having FMAs with respect to a target wire. Next, semi-redundant or redundant SMAs are removed from the rectification network. Finally, the remaining irredundant SMAs are merged and substituted by some existing nodes in the circuit. The final result achieved is equivalent to one-by-one rewiring.

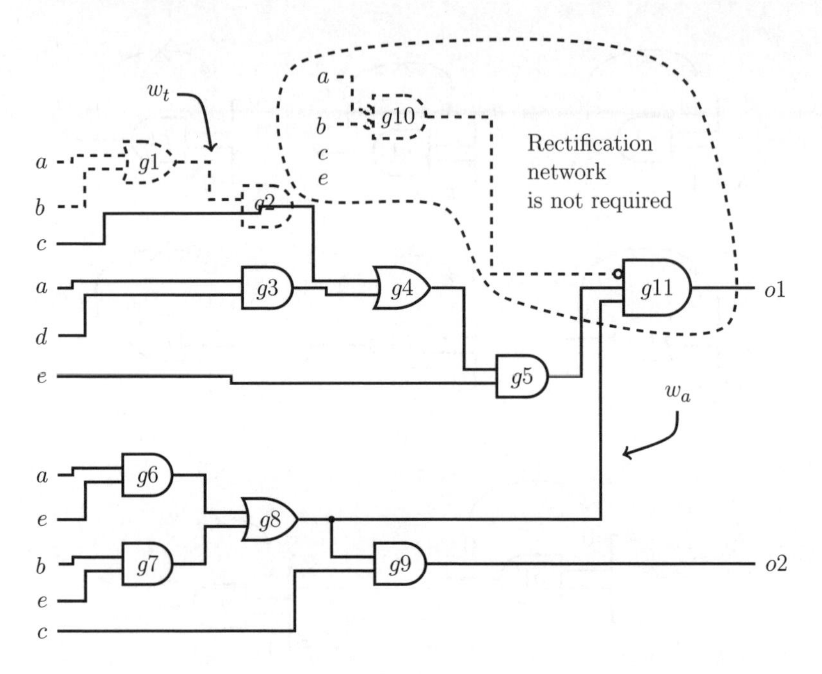

Figure 4.9 IRRA rewiring

The overall flow of the rewiring process in IRRA is listed in Algorithm 4.1. The SMAs and the Boolean network $AND(SMA_{w_t})$ can be obtained by using Algorithms 4.2 and 4.3, respectively.

4.2 ECR

Although the idea of applying a rectification network to correct the error caused by the removal of a target wire is adopted in IRRA (Lin and Wang 2009) (Section 4.1), the rewiring ability of IRRA is still similar to that of typical redundancy addition/removal (RAR)-based rewiring techniques. This is because in IRRA, the destination gate of an alternative wire has to be the dominator of the target wire or a gate with a FMA, as illustrated in Figure 4.5. The potential of error cancellation is still underexploited.

In a new study (Yang et al. 2010, Lam et al. 2012), a more general error-cancellation-based rewiring technique called ECR was proposed. The authors proved theoretically and empirically that ECR is able to locate much more alternative wires beyond the scope limited by RAR-based rewiring techniques.

ECR adheres to the same philosophy as IRRA and its rewiring process is similar to that of IRRA. It has several improvements over IRRA both in the aspects of theory and methodology.

The authors of ECR provided a deeper view on the concept of rectification networks. They derived Lemma 4.1 based on Theorem 4.1. Let $\Delta(\epsilon)$ be the *exact* set of all primary input (PI) vectors that can activate and propagate an error ϵ to primary outputs.

Algorithm 4.1: Procedure *IRRA_Rewiring*

```
    input : target wire w_t
    output: set of alternative wires W_a
 1  begin
        /* Identifying mandatory assignments                        */
 2      M ← Cal_MA(w_t);
        /* Identifying the sources of alternative wires             */
 3      SMA_{w_t} ← Cal_SMA(M);
 4      AND(SMA_{w_t}) ← Cal_AndSMA(SMA_{w_t});
 5      Classify_SMA(SMA_{w_t});
 6      Remove_Redundant_SMA(AND(SMA_{w_t}));
 7      S_S ← ∅;
 8      foreach irredundant SMA a ∈ SMA do
 9          S_a ← Find_SMA_Substitute(a);
10          S_S ← S_a;
11      S ← Find_Substitute_Intersection(S_S);
        /* Identifying the destinations of alternative wires        */
12      D ← Cal_Dominators(w_t);
13      F ← Cal_FMA(M);
14      foreach destination d in D ∪ F do
15          foreach source s in S do
16              w_a ← (s → d);
17              if d ∈ F then
18                  Verify_Alternative_Wire(w_a);
19                  if w_a is valid then
20                      W_a ← W_a ∪ w_a;
21              else
22                  W_a ← W_a ∪ w_a;
23      return W_a;
24  end
```

Algorithm 4.2: Procedure *Cal_SMA*

```
    input : set of MAs M for testing the redundancy of a target wire w_t
    output: set of source mandatory assignments SMA_{w_t}
 1  begin
 2      SMA_{w_t} = ∅ ;
 3      foreach mandatory assignment m ∈ M do
 4          if the transitive fanin cone of m contains no MAs then
 5              SMA_{w_t} ← SMA_{w_t} ∪ m;
 6      return SMA_{w_t};
 7  end
```

Algorithm 4.3: Procedure *Cal_AndSMA*

```
input  : source mandatory assignments SMA_{w_t} of a target wire w_t
output: AND(SMA_{w_t})
1 begin
2   | AND(SMA_{w_t}) = ∅ ;
3   | foreach MA m ∈ SMA_{w_t} do
4   |   └ AND(SMA_{w_t}) ← AND(SMA_{w_t}) · m ;
5   | return AND(SMA_{w_t});
6 end
```

Lemma 4.1 *Suppose there are two errors $\epsilon_r(w_t)$ and $\epsilon_a(w_a)$ in a circuit. Errors $\epsilon_r(w_t)$ and $\epsilon_a(w_a)$ cancel each other if and only if:*

1. $\Delta(\epsilon_r(w_t)) = \Delta(\epsilon_a(w_a))$, *and*
2. $\forall v \in \Delta(\epsilon_r(w_t))$, *neither D nor $\bar{D}$ can be observed at all primary outputs.*

Assume there is a vector $u \in \Delta(\epsilon_r(w_t)) \notin \Delta(\epsilon_a(w_a))$. Then, error $\epsilon_r(w_t)$ can be activated by applying u and its effect can be observed at some primary outputs and error cancellation fails under this condition.

The two errors in Lemma 4.1, namely $\epsilon_r w_t$ and $\epsilon_a(w_a)$, can be used to represent the errors caused by the removal of a target wire and the addition of an alternative wire in a circuit. In view of this, the lemma suggests that proving the validity of a potential alternative wire is in fact a problem of finding the test vectors for the errors. This problem is, in turn, a counting problem in which we are interested in how many solutions to a given problem exist, which has been proven to be NP-hard.

Actually, the rectification network for correcting the error of target wire removal can be constructed as the logical conjunction of all mandatory assignments. In addition, every test vector $u \in \Delta(\epsilon_r w_t)$ for a target wire w_t can be used to compose a rectification network. In IRRA, the intention of using SMAs to form rectification networks is not well explained. After we have discussed Lemma 4.1, the reason for that should be obvious: using dominator-based ATPG techniques to find SMAs is far less complex than finding all test vectors for a target fault. It is also desirable to derive a minimal set of mandatory assignments so that the rectification network can be as small as possible.

Using SMAs is a means to achieve the goal of constructing the simplest rectification networks with low complexity. This can further be improved as the authors of ECR have pointed out. Primarily, the definition of SMAs has a limitation. It is not general enough to cover all levels of recursive learning (Section 1.5).

Let us consider the example circuit shown in Figure 4.10. Suppose wire $g3 \rightarrow g4$ is the target wire. In order to test its stuck-at fault $sa0(g3 \rightarrow g4)$, gate $g3$ has to be assigned with value 1 to activate the fault. By recursive learning, $g3 = 1 \Rightarrow b = 1$, so the primary input b should have the value 1. For fault propagation, the primary input d should be set to 0. According to the definition of SMAs, the assignment $g3 = 1$ is not aSMA because its transitive fanin cone contains $b = 1$. Thus, only the set $\{b = 1, d = 0\}$ represents the SMAs for testing

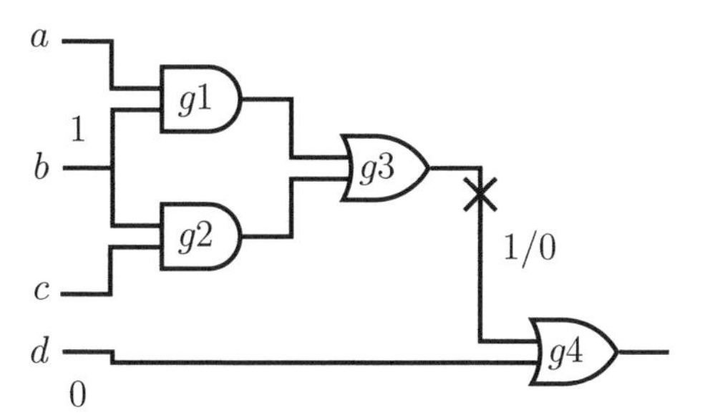

Figure 4.10 Identification of SMA under recursive learning

fault $sa0(g3 \rightarrow g4)$. However, it can be verified that setting $\{b = 1, d = 0\}$ alone cannot activate or propagate the fault. This example shows the insufficiency of the definition of SMAs, where there is a problem when the circuit contains reconvergent fanouts or under recursive learning.

Recursive learning is definitely required in logic implication and cannot be ignored. Therefore, the authors of ECR proposed a new definition to supplement SMAs.

Definition 4.4 *Given a set of mandatory assignments of a target fault, the core mandatory Assignment (CMA) is defined to be a subset of the mandatory assignments that can completely represent all other mandatory assignments.*

In other words, CMAs represent the minimum set of value assignments that cannot be forwardly implied by other mandatory assignments. Referring to the example in Figure 4.10, the set of CMAs of fault $sa0(g3 \rightarrow g4)$ is $\{b = 1, d = 0, g = 1\}$. All CMAs represent the vectors that can activate and propagate the target fault to some primary outputs.

Let $AND(CMA_{w_t})$ ($AND(CMA_{\epsilon})$) be the logical conjunction of all the CMAs for testing the removal of a target wire w_t (a error/fault ϵ). Based on the previous analysis, it can be understood that $AND(CMA_{w_t})$ represents the same condition for activating and propagating the fault caused by the addition of wire w_t's alternative wire.

Since $\Delta(\epsilon)$ represents all test vectors that can activate and propagate an error ϵ to primary outputs, we can say that it is a subset of $AND(CMA_{\epsilon})$, which is written as $\Delta(\epsilon) \subseteq AND(CMA_{\epsilon})$.

We can reformulate Lemma 4.1 according to the definition of CMAs and obtain Lemma 4.2.

Lemma 4.2 *Suppose there are two errors $\epsilon_r(w_t)$ and $\epsilon_a(w_a)$ in a circuit. Errors $\epsilon_r(w_t)$ and $\epsilon_a(w_a)$ cancel each other if and only if:*

1. *$AND(CMA_{\epsilon_r(w_t)})$ is the same as $AND(CMA_{\epsilon_a(w_a)})$, and*
2. *for any test vector in $AND(CMA_{\epsilon_r(w_t)})$, neither D nor $\bar{D}$ can be observed at all primary outputs.*

For a given error ϵ, the reason for using $AND(CMA_{\epsilon})$ instead of $\Delta(\epsilon)$ is that the former can be obtained in polynomial time and the initial rectification network can be relatively low-cost. This is a tradeoff between generality and efficiency of constructing rewiring solutions.

Implication graphs (Tafertshofer et al. 2000) can be used to compute the CMAs and $AND(CMA)$ efficiently. In this approach, amandatory assignment is represented by a node in a graph. Forward and backward implications among the mandatory assignments are represented by different kinds of edges. Then, checking whether an mandatory assignment can be implied from other mandatory assignments is simply a traversal of an implication graph. If a mandatory assignment can be forwardly implied from other nodes, it is not a CMA by definition. Since only some neighboring nodes have to be examined in practice, this kind of graph traversal for checking a mandatory assignment can always be done in linear time. The algorithm for finding CMAs is listed in Algorithm 4.4.

Algorithm 4.4: Procedure *Cal_CMA*

input : implication graph of the circuit G, set of MAs M for testing fault ϵ
output: set of core mandatory assignments CMA_ϵ

```
1 begin
2     CMA_ε = ∅ ;
3     foreach mandatory assignment m ∈ M do
4         check m in G ;
5         if m cannot be forwardly implied then
6             CMA_ε ← CMA_ε ∪ m;
7     return CMA_ε;
8 end
```

4.2.1 Destination of Alternative Wires

The concept of dynamic dominators is a major idea on which the theories in ECR are based. It is defined as follows:

Definition 4.5 *During a stuck-at fault test, a node in the circuit that is necessary for the fault to propagate to primary outputs is called a dynamic dominator.*

Definition 4.6 *Given a network, a node set S is a blocking cut set of a wire w if every path from wire w to primary outputs must go through one and only one node in S.*

Dynamic dominators were first explained in Krieger et al. (1991). As we can see from the definition, dynamic dominators are not the same as the dominators that were been introduced in Section 1.4. Dominators of a gate or a wire can be identified structurally using Algorithm 1.1, whereas dynamic dominators of a fault are not based on the circuit structure alone. Dynamic dominators of a fault are determined according to the test vectors that can test the fault and the corresponding fault propagation. A dynamic dominator is a special node in a *blocking cut set*.

In Section 2.2.2.2, we have already used an example circuit to explain the concepts of dynamic dominators (Figure 2.10). Another more complex example circuit is given in Figure 4.11. Wire $a \rightarrow g1$ is the target wire, and the corresponding fault to test is $sa1(a \rightarrow g1)$. Gates $g1$, $g3$, and $g8$ are common in the paths from the wire to the primary output at gate $g8$. Thus, they are dominators of the fault. Dominators are indicated by dotted

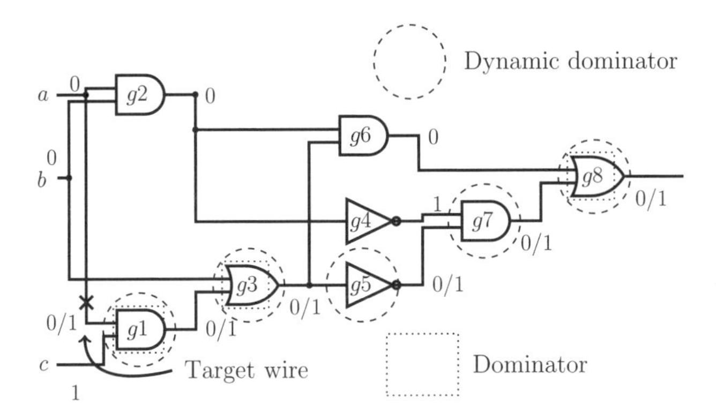

Figure 4.11 Examples of dynamic dominators

rectangles in the figure. By tracing the path of fault propagation, it can be found that gate $g1$ is also a dynamic dominator of the fault. Besides gate $g1$, gates $g3$, $g5$, $g7$, and $g8$ are also the dynamic dominators. They are indicated by dashed circles. There is a branch at the output of gate $g3$. However, not all fanouts of gate $g3$ are dynamic dominators. The fault cannot propagate through gate 6 because one of its input has a controlling value. As can be seen from this example, we know that dynamic dominators may be more abundant than dominators. If they can be applied as the destination nodes of alternative wires, the rewiring ability can certainly be improved.

The dynamic dominators of a fault is similar to the structural dominators in that they all have either D or $\bar{D}$ values. They can be used as the destination nodes of alternative wires by canceling the errors that pass through them. Some of the error cancellation rules for basic gate types are shown in Figure 4.12. For example, if D and $\bar{D}$ are two inputs to an AND gate, these errors cancel each other and the output of the AND gate becomes a constant 0. The values of other inputs do not matter. An XOR gate outputs 0 when the values of both its inputs are the same, and outputs 1 when the values of both its inputs are different. Therefore, the errors are cancelled in the cases where the inputs of an XOR gate have values D and D, $\bar{D}$ and $\bar{D}$, or D and $\bar{D}$.

Dynamic dominators are further classified as described in Definition 4.7.

Definition 4.7 *A node d is a $sa1$ dynamic dominator of the $sa1(w)$ fault of a wire w if the stuck-at fault must go through d so as to be propagated to primary outputs.*

A node d is an $sa0$ dynamic dominator of the $sa0(w)$ fault of a wire w if the stuck-at fault must go through d so as to be propagated to primary outputs.

In the rewiring scheme, whether the stuck-at fault for the target wire is s-a-0 or s-a-1 fault is determined by the polarity and the sink node of the target wire. In the following, we will use the simplified notation *dynamic dominator* if it is clear from the context.

Figure 4.13 shows an example of *dynamic dominator*: Regarding $sa0(b \rightarrow g1)$, the fault can be propagated through gates $g7$ or $g8$. For fault activation, the primary input b must be assigned to 1. However, $b = 1$ implies $g6 = 0$ and the path through $g8$ is blocked. As a result, the fault can be propagated only though $g7$. Thus, gate $g7$ is a *dynamic dominator* of the fault $sa0(b \rightarrow g1)$.

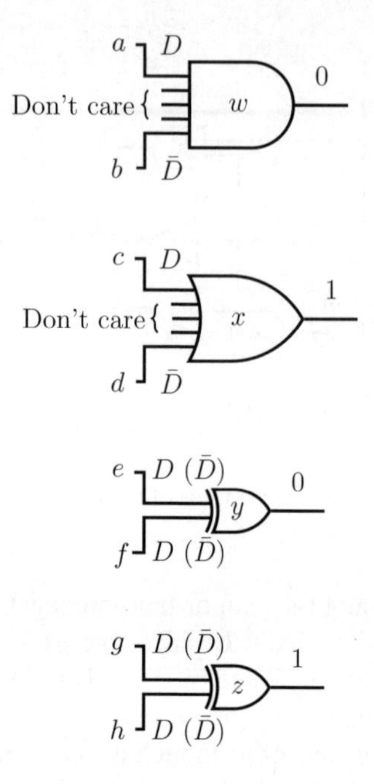

Figure 4.12 Examples of error cancellation rules

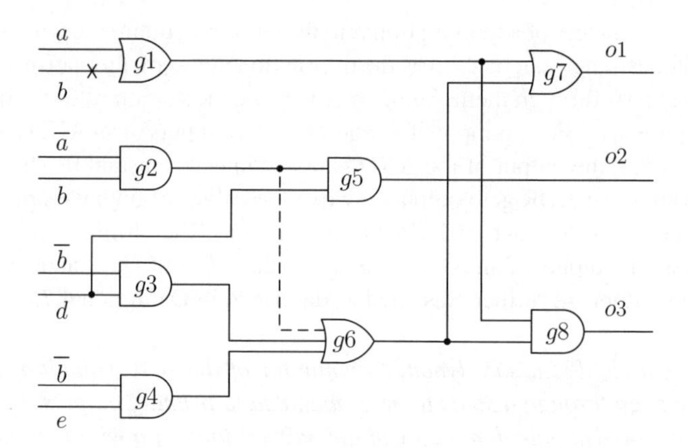

Figure 4.13 Generalized wire addition and removal

4.2.1.1 Dynamic Dominator as Destination

If the destination node n_d of an alternative wire is a dynamic dominator, the error injected by removing the target wire ($\epsilon_r(w_t)$) has an opportunity to be cancelled completely at this node. The value of $AND(CMA)$ is 1 when $\epsilon_r(w_t)$ is activated and propagated.

Theorem 4.5 *A node n_d is a valid destination of the rectification network ($AND(CMA)$) for the removal of an irredundant target wire w_t if n_d is a dynamic dominator with an MA for the corresponding stuck-at fault on the wire w_t.*

Proof. When n_d is a dynamic dominator, it has an MA whose value is either D(1/0) or $\bar{D}$(0/1) because the error can be propagated to one of the primary outputs. Suppose the MA of the dynamic dominator n_d is D(1/0). As $AND(CMA)$ has a logic value 0 when the error $\epsilon_r(w_t)$ is not activated or propagated and 1 otherwise, $n_d + AND(CMA)$ will assume a constant logic 1. In other words, the error $\epsilon_r(w_t)$ is cancelled. Similarly, when the MA of n_d is $\bar{D}$(0/1), we can use $n_d \cdot \overline{AND(CMA)}$ to cancel the error. ■

For example, in Figure 4.14, suppose the target wire w_t is $b \rightarrow g1$. The gate $g7$ is a dynamic dominator with respect to w_t and has an MA value of D. In addition, the set of CMA is $\{a = 0, b = 1\}$. The rectification network at $g7$ is indicated by the dotted box. The wire removal error $\epsilon_r(b \rightarrow g1)$ is finally rectified at $g10$, which is added directly at the output of the dynamic dominator $g7$.

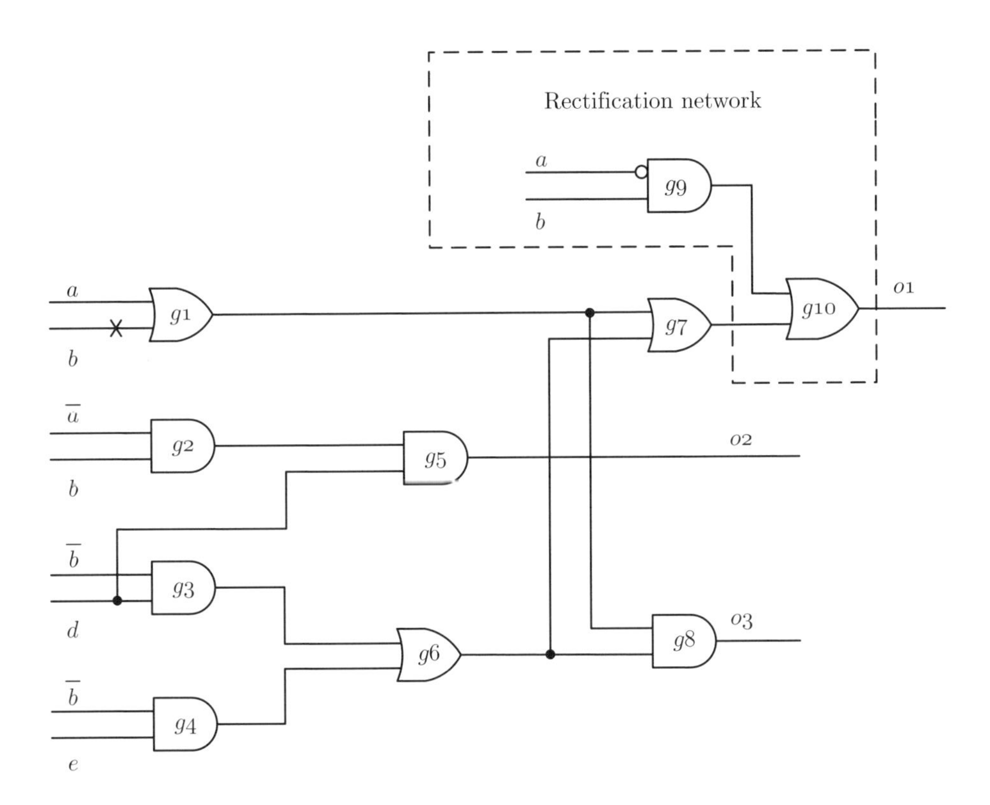

Figure 4.14 Rectification network

4.2.1.2 Node with FMA as Destination

If the destination node n_d is not a dynamic dominator but is a node with FMA instead, the error injected by wire removal error $\epsilon_r(w_t)$ can have a chance to be cancelled at the transitive fanout cone of the target wire.

However, changing the value of the node with FMA is not always safe because this operation may cause some new errors. If the wire addition error $\epsilon_a(w_a)$ injected at a node n_d cannot be cancelled by the wire removal error $\epsilon_r(w_t)$, n_d cannot not be a destination for the alternative wire. Therefore we have to verify whether the error effect is observable when both errors $\epsilon_a(AND(CMA) \rightarrow n_d)$ and $\epsilon_r(w_t)$ are injected into the circuit before taking the node n_d as a valid destination of the rectification network ($AND(CMA)$).

From the previous analysis on CMA, we know that $AND(CMA)$ represents all the test vectors that can activate and propagate the wire removal error $\epsilon_r(w_t)$. Rectification networks $n_d + AND(CMA)$ and $n_d \cdot \overline{AND(CMA)}$ can also be utilized to make sure the wire addition error $\epsilon_a(AND(CMA) \rightarrow n_d)$ is only activated and propagated under the same test vectors. For a node n_d with MA 0 or D, we can use $n_d + AND(CMA)$ to replace n_d. The value of $n_d + AND(CMA) = n_d$ when $\epsilon_r(w_t)$ is not activated or propagated. When $\epsilon_r(w_t)$ is injected, $AND(CMA) = 1$ and $n_d = 0$ in the faulty circuit. So, the value of $n_d + AND(CMA) \neq n_d$. If n_d has MA 1 or $\overline{D}$, we can use $n_d \cdot \overline{AND(CMA)}$ to generate the wire addition error $\epsilon_a(\overline{AND(CMA)} \rightarrow n_d)$. When $\epsilon_r(w_t)$ is not generated, $\overline{AND(CMA)} = 1$, so $n_d \cdot \overline{AND(CMA)} \equiv n_d$. Both errors $\epsilon_r(w_t)$ and $\epsilon_a(\overline{AND(CMA)} \rightarrow n_d)$ are not generated. When $\epsilon_r(w_t)$ is activated and propagated, $\overline{AND(CMA)} = 0$. The value of n_d will be changed in the faulty circuit.

After both errors are injected into the circuit, the value of CMAs, which is the representation of the set of test vectors, should also be injected into the circuit to propagate the errors. After logic implications, all the nodes with value $0/0$ or $1/1$ are collected in set M. If all the paths

---

Algorithm 4.5: DestinationCheck(Circuit C, target wire w_t, CMA set S, candidate destination n_d)

---

input : Circuit C, target wire w_t, CMA set S, candidate destination n_d
output: whether the destination is valid

```
1  begin
2      foreach node s in S do
3          assign MA of s;
4      inject ε_r(w_t) into C;
5      if MA of n_d is 0 or D then
6          inject n_d s-a-1 fault into C ;
7      else
8          inject n_d s-a-0 fault into C ;
9      launch error propagation in C;
10     if both ε_r(w_t) and ε_a(n_d) cannot propagate to PO then
11         n_d is a valid destination;
12 end
```

from either error to POs must go through M, these two errors cancel each other before they reach primary outputs. Thus node n_d is a valid destination node of the alternative wire.

Referring to the example in Figure 4.14 again, in order to propagate the fault $sa0(b \rightarrow g1)$, the side inputs of all dynamic dominators should be assigned to their noncontrolling values. An FMA 0 is thus assigned to both $g6$ and a. The MAs are $\{a = 0, b = 1, g1 = D, g2 = 1, g3 = 0, g4 = 0, g6 = 0, g7 = D, g8 = 0\}$, and CMAs are $\{a = 0, b = 1\}$. To verify whether $g6$ is a feasible destination, both faults $sa0(b \rightarrow g1)$ and $sa1(g6)$ are injected into the circuit. At the same time, the values of CMAs are assigned too. After implications, it is found that $g7 = 1$ and $g8 = 0/0$, which implies that the function of the circuit is unchanged when both errors are injected, and therefore node $g6$ is a valid destination node for alternative wires. The dotted line shows one possible alternative wire $g2 \rightarrow g6$.

4.2.2 *Source of Alternative Wires*

Given that the errors cancel each other at g_d, and suppose that the function of g_d in the original circuit is denoted as F and the function in the rewired circuit (after adding rectification network) is denoted as F'.

$$F \oplus F' = 0$$

This equation shows that F and F' are strictly equivalent. However, more flexibility could be gained in rewiring if we relax it to

$$F \oplus F' \subseteq ODC(g_d)$$

where ODC is the function calculating the observability don't cares (Damiani and De Micheli 1990) of a Boolean variable. Those are the conditions under which the variable is not affecting any primary outputs. The relaxation is valid because all conditions causing F and F' to be different at g_d will not be observable at any primary output.

Algorithm 4.6: SourceNodeIdentification(Circuit $C + \Gamma$, target wire w_t, node g_{CMA})

```
   input  : Circuit C + Γ, target wire w_t, node g_CMA
   output: SourceNode S
 1 begin
 2   SourceNode S = ∅ ;
 3   remove w_t from C + Γ ;
 4   compute MA_0 for g_CMA s-a-0 test ;
 5   compute MA_1 for g_CMA s-a-1 test ;
 6   foreach node n having different values in MA_0 and MA_1 do
 7     if n is not transitive fanout of g_CMA then
 8       insert n into S ;
 9   return S;
10 end
```

A rectification network with root node g_{CMA} is denoted as Γ. According to the previous analysis, $C \equiv C - w_t + \Gamma$. Since Γ may not be a single gate, substitution nodes for g_{CMA} have to be determined in order to transform Γ into a single alternative wire.

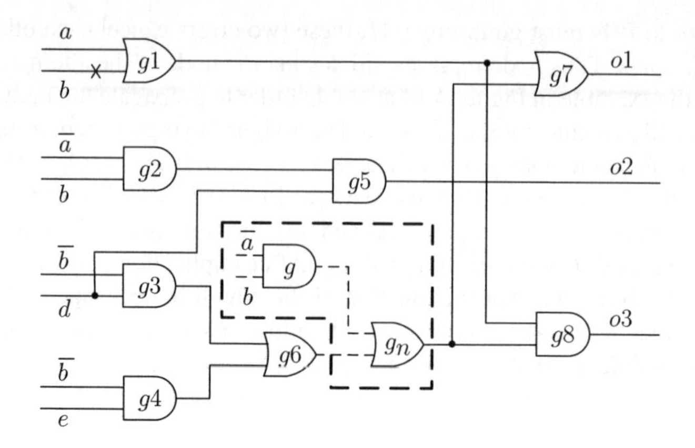

Figure 4.15 Destination identification

The authors in Chen and Wang (2009) proposed an algorithm that requires only two stuck-at fault tests to find the substitution nodes for a given node. The nodes that have different values in the stuck-at fault tests for $(sa0(g_{CMA}))$ and $(sa1(g_{CMA}))$ are out of the transitive fanout cone of g_{CMA} and can be valid substitution nodes for g_{CMA}. These two values are denoted by MA_0 and MA_1, respectively.

For example, consider finding substitution nodes for the rectification network shown in Figure 4.15. Suppose that the recursive learning depth is 1 when the stuck-at fault is tested. First, we compute MAs for the $sa0$ test on g_{CMA}. To activate this fault, g_{CMA} is set to D. To propagate the fault, $g6$ is set to 0. Then logic implications are performed to find MAs. The set of MAs for the test for $sa0(g_{CMA})$ is $\{a = 0$, $b = 1$, $g2 = 1$, $g3 = 0$, $g4 = 0$, $g6 = 0$, $g_{CMA} = D$, $g_n = D$, $g7 = D$, and $g8 = 0\}$. Second, we use the same method to calculate the MAs for the test for $sa1(g_{CMA})$. The set of MAs is $\{g2 = 0$, $g3 = 0$, $g4$=0, $g5$=0, $g6$=0, $g_{CMA} = \bar{D}$, and $g_n = \bar{D}\}$. Finally, it is found that $g2$ has different values in the two sets of MAs. Hence, $g2$ is a source node corresponding to the destination $g6$.

4.2.3 *Overview of the Approach of Error-Cancellation-Based Rewiring*

In this approach, a candidate destination for an alternative wire is located first, and then the new error induced at the candidate destination is checked to see whether it can cancel the error injected by the removal of the target wire. Finally, the source nodes are found for all the destinations to give the set of alternative wires.

The proposed algorithm, whose pseudo-code is shown in Algorithm 4.7, is listed as follows. First, the corresponding stuck-at fault is assumed and processed on w_t to find the MAs, CMAs, and candidate destinations. Second, for each candidate destination, both the wire removal error $\epsilon_r(w_t)$ and the wire addition error $\epsilon_a(w_a)$ are simultaneously inserted into the circuit to verify whether the destination is valid. Finally, for each valid destination node, the source node is identified to form a single alternative wire w_a.

Algorithm 4.7: FindAlternativeWire(Circuit C, target wire w_t)

```
   input  : Circuit C, target wire w_t
   output: alternative wires AW
1  begin
2      AW = ∅;
3      launch stuck-at fault test on w_t, collect dynamic dominators(see Section 2.2.2.2)
       and MAs as candidate destinations (D) ;
4      foreach each candidate destination node d in D do
5          if ε_a(c_d) and ε_r(w_t) cancel each other(see Section 4.2.1) then
6              insert d into D' ;
7      foreach each destination node d in D' do
8          find source node set S(see Section 4.2.2);
9          foreach source node s in S do
10             insert (s, d) into AW;
11     return AW;
12 end
```

4.2.4 *Complexity Analysis of ECR*

We assume that the time complexity for a stuck-at fault test is no more than a constant T and the number of nodes in a circuit is n. The number of target wires is close to $2n$ if a circuit is composed of only two-input gates.

In the process of IRRA, MAs, and CMAs (SMAs as in the paper) are calculated by a stuck-at fault test for a target wire. It costs T to launch the stuck-at fault test. It costs $\mid FMA \mid \cdot T$ to determine the valid destinations. In the substitution procedure, the time complexity is no more than $\mid irredundant\ CMA \mid \cdot O(n)$. In the worst case, the numbers of FMAs and irredundant CMAs are no more than n. Therefore, the time complexity for IRRA is $O(n^2)$. In practice, the number of irredundant CMAs is not larger than a certain constant; therefore, it normally takes nearly linear time to find alternative wires for a given target wire. Table 4.1 shows the experimental results for IRRA on CPU time. According to the previous analysis, the relationship between CPU time and the number of irredundant CMAs are calculated. The first column shows the name of the benchmarks. The second column gives the number of target wires in the benchmark. The third column lists the CPU time cost. The average CPU time cost for each target wire is denoted by avg.TIME. |ir *CMA*| is the number of irredundant CMAs for all target wires. avg. |ir *CMA*| is the average number of irredundant CMAs for each target wire. The last column presents the quotient: avg.TIME divided by avg. |ir *CMA*|. The relation between circuit size and avg.TIME/avg. |ir *CMA*| is expressed by a scatter plot as in Figure 4.16. As shown in this figure, avg. TIME/avg.|ir *CMA*| is close to linear with circuit size.

The authors of ECR implemented the rewiring algorithm in a such way that dynamic dominators are determined by analyzing the blocking cut sets near the regular dominators. The time complexity of finding the dynamic dominator is $T \cdot |\{dominators\}|$. To determine whether a

Table 4.1 CPU time analysis for IRRA

Benchmark	$\|TW\|$	TIME	avg.TIME	$\|\text{ir } CMA\|$	avg.$\|\text{ir } CMA\|$	$\frac{\text{avg.TIME}}{\text{avg.}\|\text{ir } CMA\|}$
9sym-hdl	96	0.99	0.010	394	4.10	0.0025
pcler8	126	4.1	0.033	1,350	10.71	0.0030
f51m	190	8.5	0.045	2,189	11.52	0.0038
comp	184	28.2	0.153	3,788	20.59	0.0074
5xp1	178	6.86	0.039	1,918	10.78	0.0036
b9_n2	166	5.24	0.032	1,147	6.91	0.0046
my_adder	256	6.27	0.024	796	3.11	0.0079
ttt2	274	13.85	0.051	2,058	7.52	0.0067
term1	314	32.28	0.103	4,246	13.52	0.0076
sao2-hdl	324	50.91	0.157	5,056	15.60	0.0101
C432	320	43.97	0.137	4,299	13.43	0.0102
duke2	566	144.44	0.244	9,068	16.02	0.0159
alu2	662	276.02	0.417	12,163	18.37	0.0227
C1908	706	148.37	0.210	6,743	9.55	0.0220
misex3	840	365.32	0.435	13,325	15.86	0.0274
C880	648	124.52	0.192	8,186	12.63	0.0152
C1355	772	124.92	0.162	5,084	6.58	0.0246
rot	942	235.97	0.250	7,322	7.77	0.0322
x3	1,120	435.27	0.389	11,911	10.63	0.0365
apex6	1,188	494.90	0.417	12,437	10.47	0.0398
C499	772	126.66	0.164	5,084	6.59	0.0249
alu4	1,252	1206.01	0.963	34,709	27.72	0.0347
C2670	1,074	1081.55	1.007	34,634	32.25	0.0312
C3540	1,870	2754.63	1.473	56,587	30.26	0.0487
C5315	2,622	2094.18	0.799	21,847	8.33	0.0959
C7552	2,876	3956.63	1.376	43,996	15.30	0.0899

$|TW|$: the number of target wires.
TIME: The CPU time in seconds.
avg.TIME: TIME/$|TW|$.
$|\text{ir}CMA|$: The number of irredundant CMAs.
avg.$|\text{ir}CMA|$: $|\text{ir}CMA|/|TW|$.

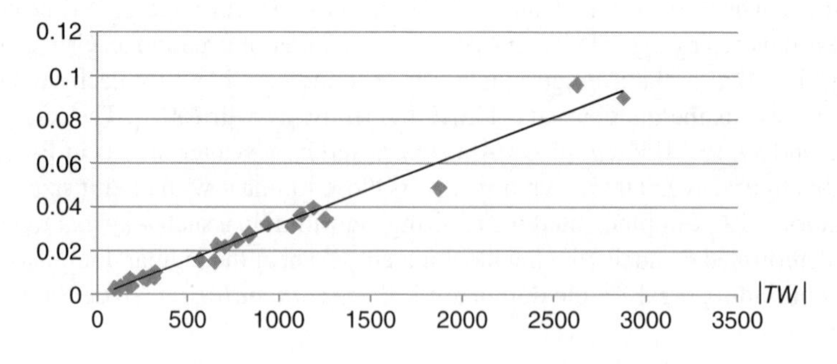

Figure 4.16 Empirical runtime of IRRA

candidate destination c_d is a valid destination, the complexity of the propagation of two errors and justification is no more than $O(n)$. The number of candidate destination nodes is $\mid CD \mid$, so the time consumed for destination identification is no more than $\mid CD \mid \cdot O(n)$. For each destination node in D, two stuck-at fault tests are performed to find the source nodes. The time consumed for source node identification is no more than $2T\cdot \mid D \mid$. The time complexity of the ECR algorithm is $T\cdot \mid \{dominators\} \mid + \mid CD \mid \cdot O(n) + 2T\cdot \mid D \mid$. As the number of dominators and candidate destinations is no more than n, the time complexity of the ECR algorithm is $O(n^2)$ in the worst case.

Table 4.2 CPU time analysis for ECR

Benchmark	$\|TW\|$	$\|dom\|$	D	CD	avg. $\|dom\|$	avg. D	avg. CD	TIME	avg. TIME	$\frac{\text{avg.TIME}}{\text{avg.}CD}$
9sym-hdl	96	476	184	417	4.96	1.92	4.34	0.99	0.010	0.0024
pcler8	126	464	372	1,160	3.68	2.95	9.21	4.1	0.033	0.0035
f51m	190	778	787	1,336	4.09	4.14	7.03	8.5	0.044	0.0064
comp	184	456	1,772	3,036	2.48	9.63	16.5	28.2	0.153	0.0093
5xp1	178	702	708	1,138	3.94	3.98	6.39	6.86	0.039	0.0060
b9_n2	166	502	508	991	3.02	3.06	5.97	5.24	0.032	0.0053
my_adder	256	520	416	1,048	2.03	1.63	4.09	6.27	0.024	0.0060
ttt2	274	886	832	1,690	3.23	3.04	6.17	13.85	0.051	0.0082
term1	314	1,282	1,536	3,099	4.08	4.89	9.87	32.28	0.103	0.0104
sao2-hdl	324	988	1,534	3,791	3.05	4.73	11.70	50.91	0.157	0.0134
C432	320	996	1,355	3,149	3.11	4.23	9.84	43.97	0.137	0.0140
duke2	566	2,174	2,331	6,342	3.84	4.12	11.20	144.44	0.255	0.0228
alu2	662	3,326	3,219	6,693	5.02	4.86	10.11	276.02	0.417	0.0412
C1908	706	1,952	1,882	6,370	2.76	2.67	9.02	148.37	0.210	0.0233
misex3	840	3,404	3,759	8,781	4.05	4.48	10.45	365.32	0.435	0.0416
C880	648	2,734	2,548	4,934	4.22	3.93	7.61	124.52	0.192	0.0252
C1355	772	2,332	1,472	4,984	3.02	1.91	6.46	124.04	0.161	0.0249
rot	942	2,840	2,899	5,903	3.01	3.08	6.27	227.36	0.241	0.0385
x3	1,120	3,976	4,168	8,346	3.55	3.72	7.45	405.82	0.362	0.0486
apex6	1,188	4,548	4,463	9,581	3.83	3.77	8.06	469.82	0.395	0.0490
C499	772	2,380	1,472	4,984	3.08	1.91	6.46	122.96	0.159	0.0247
alu4	1,252	5,908	8,397	16,998	4.72	6.71	13.58	1,661.69	1.327	0.0978
C2670	1,074	5,278	8,437	15,192	4.91	7.86	14.15	958.83	0.893	0.0631
C3540	1,870	7,850	12,553	27,549	4.20	6.71	14.73	3,290.54	1.760	0.1194
C5315	2,622	9,516	8,423	18,433	3.63	3.21	7.03	1,998.72	0.762	0.1084
C7552	2,876	12,442	11,540	29,891	4.33	4.01	10.39	4,024.5	1.399	0.1346

D: Number of valid destination nodes.
CD: The number of candidate destination nodes.
avg.|*dom*|: |*dom*|/|*TW*|.
avg.*D*: *D*/|*TW*|.
avg.*CD*: *CD*/|*TW*|.
TIME: CPU time in seconds.
avg.TIME: TIME/|*TW*|.

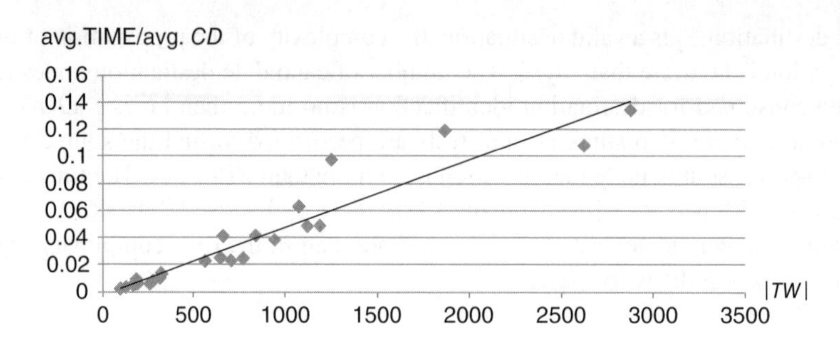

Figure 4.17 Empirical runtime of ECR

The statistical results of the analysis on the data $\mid \{dominators\} \mid$, $\mid CD \mid$, and $\mid D \mid$ show that their values are generally small and are not dependent on the circuit size n in practice. In Table 4.2, the third column shows the number of dominators for all target wires.

D is the number of valid destination nodes of all the alternative wires for all the target wires. CD is the number of candidate destination nodes for all the potential alternative wires. avg. $|dom|$, avg. D, and avg. CD are the average number of dominators, valid, and candidate destination nodes for each target wire. TIME gives the CPU time cost for all target wires. avg. TIME is the average CPU time cost for each target wire. The last column represents the ratio of avg. TIME to avg. CD. As illustrated in Figure 4.17, nearly $O(n)$ time is required to find alternative wires for a given target wire.

4.2.5 Comparison Between ECR and Other Resynthesis Techniques

4.2.5.1 Comparison Between ECR and Traditional Boolean Network Optimizations

Rewiring is different from traditional Boolean network optimizations using algebraic techniques and those using Boolean techniques such as don't care-based optimizations. In general, rewiring is an ATPG-based logic optimization algorithm that exploits heavily both structural and logic information of a circuit.

ECR and other rewiring schemes differ in how and where a target wire can be made redundant and what transformations should be added to recover the incorrect circuit function created intentionally.

The solution space of different optimization methods, however, can overlap. Comparisons between ECR and several selected optimizations methods are further discussed in the following sections.

4.2.5.2 Comparison Between ECR and Rewriting

Rewriting (Mishchenko et al. 2006) is a greedy algorithm for technology-independent logic synthesis, which can be used to minimize the and-inverter graph (AIG) size by iteratively replacing AIG subgraphs rooted at a node by smaller precomputed subgraphs. This resynthesis technique is fast and can be applied to a circuit many times. Its quality of resynthesis depends on the availability and quality of the precomputed subgraphs. Typically, it enumerates all 4-feasible cuts of a root node to find the most appropriate subgraph. Since cut enumeration for larger cut sizes is not practical in general, AIG rewriting is a pretty local resynthesis technique. The functions of the root nodes before and after rewriting are the same because rewriting does not explore the use of observability don't cares and other kinds of don't cares. The fanouts of the root nodes after rewriting are also the same as before.

On the other hand, ECR is a global resynthesis method whose idea is to cancel any errors before they are propagated to the primary outputs of the circuit. There is no constraint on the location of the alternative wire for a given target wire as long as they can cancel each other. Although ECR is a method for wire replacement, the effect of rewiring is sometimes the same as rewriting. If the sink nodes of the target wire and the alternative wire are the same, the functions of the target wire's source node and the alternative wire's source node may also be the same. If this is the case, the effect of rewiring is the same as rewriting. Generally, the functions of the target wire's source node and the alternative wire's source node are different in ECR.

4.2.5.3 Comparison Between ECR and Node Merger

Node merger is an efficient logic restructuring method. It replaces a node in the circuit with another node that can be a new node or an existing node in the circuit without changing the functionality of the circuit. Two nodes can be merged if they are functionally equivalent or their functional differences cannot be observed at any primary outputs. There are two kinds of node merging techniques: SAT-based approaches (Zhu et al. 2006, Plaza et al. 2007) and ATPG-based approaches (Chen and Wang 2009, 2010a, b, 2012). The solution spaces of node merger and ECR overlap when the source node of the target wire has only one fanout and the alternative wire's sink node is also the sink node of the target wire.

4.2.5.4 RAR Is a Special Case of ECR

In the process of RAR, the mandatory assignments and FMAs for the stuck-at fault test of a given target wire $w_t(n_a \rightarrow n_b)$ in a circuit C are first calculated so as to find the alternative wires for the target wire. The stuck-at fault of the target wire can be blocked if there are some MA values conflicting with the FMA values. The conflicts always occur at w_t's dominators with side inputs. A conflict in mandatory assignments can be constructed by adding a candidate alternative wire w_a. This means that w_t is redundant in the network $C + w_a$ and the construction satisfies that $C + w_a \equiv C - w_t + w_a$. In addition, the redundancy of w_a in $C + w_a$ is checked. If w_a is redundant and $C + w_a \equiv C$, it can be concluded that adding the redundant wire w_a makes the target wire redundant and w_a is a valid alternative wire for the target wire.

Because the MAs in RAR are calculated by the side input assignments of the dominators, the $\epsilon_r(w_t)$ is always corrected at the dominators. However, the MAs in ECR are calculated by setting the side inputs of the dynamic dominators. The wire removal error $\epsilon_r(w_t)$ can be cancelled at any blocking cut set. As shown in Figure 4.15, the wire removal error $\epsilon_r(w_t)$ and the wire addition error $\epsilon_a(w_a)$ cancel each other at a blocking cut set $\{g7, g8\}$. Moreover, ECR uses error cancellation to identify destinations of alternative wires. The errors $\epsilon_r(w_t)$ and $\epsilon_a(w_a)$ may cancel each other at locations that cannot be identified by RAR.

As illustrated in the example in Figure 4.18 (Chang and Marek-Sadowska 2007), the node $g7$ is a dynamic dominator of the stuck-at fault $sa1(a \rightarrow g1)$. The MAs to activate and propagate

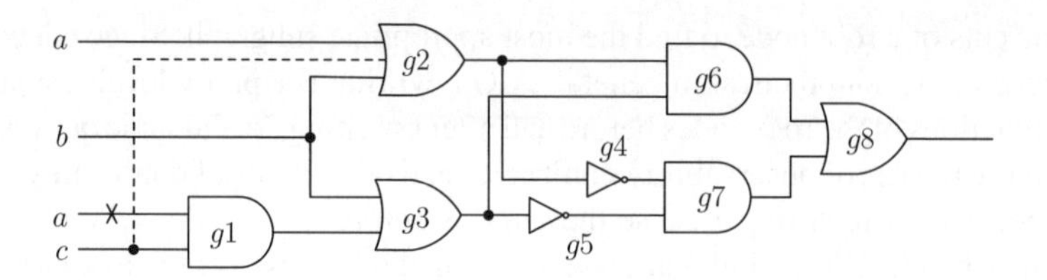

Figure 4.18 An error cancelled at dominator that does not have side input

$sa1(a \to g1)$ are: $\{a = 0, c = 1, g1 = \bar{D}, b = 0, g3 = \bar{D}, g2 = 0, g5 = D, g4 = 1, g6 = 0, g7 = D$, and $g8 = D\}$. Adding the wire $c \to g2$ creates conflicts in the mandatory assignments of $g2$. However, the additional wire $c \to g2$ is irredundant in the circuit $C + (c \to g2)$. If we activate and propagate the errors $\epsilon_r(a \to g1)$ and $\epsilon_a(c \to g2)$ simultaneously, these two errors cancel each other before they reach primary outputs. First, the values of the SMAs for testing $sa1(w_t)$ are set in the circuit. The SMAs are $\{a = 0, b = 0, c = 1\}$. Second, $\epsilon_r(a \to g1)$ $(\bar{D})$ and $\epsilon_a(g2)$ $(\bar{D})$ are inserted into circuit because $g2$ has an MA 0 for observing the error $\epsilon_r(a \to g1)$. After implications, $g6 = 0/1$ and $g7 = 1/0$. Since $g8$ is an OR gate, the value of $g8$ is 1/1. These two errors cancel each other at the dominator $g8$. So, node $g2$ is a destination node for the alternative wire. After the process of source identification, c is found to be a valid source node with regard to $g8$. Therefore, $c \to g2$ is an alternative wire for the target wire $a \to g1$.

4.2.6 Experimental Result

The authors of Lam et al. (2012) compared IRRA and ECR empirically. They compared the rewiring capability of the two rewiring algorithms by (i) comparing the number of alternative wires identified in a circuit and (ii) comparing the quality of alternative wires by applying them to logic transformations to improve technology mapping.

4.2.6.1 Rewiring Ability

A set of combinational circuits and a set of sequential circuits were adopted to test the rewiring ability of the rewiring algorithms. The benchmark sets were first processed by `resyn2` in ABC (Berkeley Logic Synthesis and Verification Group) and then decomposed into simple gates by `tech_decomp; dmig` in RASP SIS (Sentovich et al. 1992).

Table 4.3 shows the result of single alternative wire identification. $|TW|$ represents the number of target wires in each circuit. The third column shows the percentage of target wires having alternative wires in IRRA. The fourth column lists the number of alternative wires found by IRRA. The fifth column presents the CPU time measured in seconds. The percentage of target wires having alternative wires in ECR is listed in the sixth column. The seventh column gives the number of alternative wires found by ECR. The eighth column gives the percentage of alternative wires by ECR that cannot be found by IRRA. The last column lists the CPU time. In this experiment, ECR identified 76% more target wires that had alternative wires. Moreover, 56% of the alternative wires by ECR could not be found by IRRA. The total number of alternative wires found by ECR was nearly twice (202%) that found by IRRA. The CPU time penalty was only 8%.

Table 4.3 Comparison on combinational benchmarks between IRRA and ECR

Benchmark	$\|TW\|$	IRRA			ECR			
		%	$\|AW\|$	TIME	%	$\|AW\|$	% non-IRRA	TIME
9sym-hdl	96	6.25	10	0.68	26.04	58	82.8	0.99
pcler8	126	15.87	37	3.41	28.57	106	62.3	4.1
f51m	190	30.53	161	7.62	49.47	441	65.5	8.5
comp	184	21.74	187	13.92	42.39	610	74.0	28.2
5xp1	178	32.58	197	6.11	50.56	452	58.7	6.86
b9_n2	166	33.73	149	5.13	46.99	240	58.7	5.24
my_adder	256	12.50	82	5.83	31.64	134	38.8	6.27
ttt2	274	29.93	340	12.52	43.80	521	41.4	13.85
term1	314	29.94	373	32.18	45.22	699	57.5	32.28
sao2-hdl	324	17.59	149	38.21	29.63	428	55.1	50.91
C432	320	26.25	315	35.05	40.94	545	56.7	43.97
duke2	566	33.92	1,274	126.71	44.35	2,230	48.2	144.44
alu2	662	29.61	861	182.10	47.89	2,749	72.7	276.02
C1908	706	8.36	121	141.13	22.95	291	59.7	148.37
misex3	840	30.24	1,860	271.14	41.31	3,170	50.4	365.32
C880	648	16.20	198	119.02	29.78	458	61.1	124.52
C1355	772	1.55	68	124.92	22.28	228	70.2	124.04
rot	942	26.75	618	235.97	42.25	1,046	41.3	227.36
x3	1,120	17.59	759	435.27	30.45	1,287	41.3	405.82
apex6	1,188	21.13	1,219	494.9	31.90	2,099	43.9	469.82
alu4	1,252	15.73	1,931	1,206.01	49.84	3,826	55.4	1,661.69
C2670	1,074	16.57	1,266	1,081.55	29.98	2,960	58.1	958.83
C3540	1,870	21.34	2,399	2,754.63	36.04	4,051	42.0	3,290.54
C5315	2,622	14.07	1,021	2,094.18	28.34	2,542	56.8	1,998.72
C7552	2,876	12.13	857	3,956.63	24.86	2,039	59.2	4,024.5
Total	19,566		16,452	13,384.82		33,210		14,421.16
Average		20.88			36.70		56.0	
Ratio					1.76	2.02		1.08

$|TW|$: Number of target wires.
%: Percentage of target wires having alternative wires.
$|AW|$: Number of alternative wires.
% non-IRRA: Percentage of alternative wires by ECR that cannot be identified by IRRA.
TIME: The CPU time, measured in seconds.

Table 4.4 lists the experiment on sequential circuits. In this experiment, ECR obtained 17% improvement on the percentage of target wires that had alternative wires. ECR could identify about 23% more total number of alternative wires than IRRA. The penalty on CPU time was 22%. This is reasonable because the combinational part of the circuit is relatively small in sequential circuits.

For several relatively large benchmarks, ECR required smaller computation time than IRRA. It is believed that this is because IRRA has to determine the redundancy of each of the MAs constituting a rectification network; and since a target wire in a larger circuit may have a higher chance to have more MAs, more computation effort is required.

Table 4.4 Comparison on sequential benchmarks between IRRA and ECR

Benchmark	|*TW*|	IRRA			ECR		
		%	|*AW*|	TIME	%	|*AW*|	TIME
s27	88	84.09	343	3.00	88.64	372	3.79
sao2	276	57.61	923	24.66	59.78	949	31.43
s208	120	63.33	217	6.67	80.00	332	7.06
s208_1	110	80.91	276	8.21	85.45	292	10.37
sct	244	71.72	910	24.39	78.69	967	29.78
s420	178	76.97	579	17.05	85.96	628	20.31
s641	298	65.10	880	49.23	78.19	1,212	61.66
s349	240	57.92	498	29.05	72.08	630	34.05
s344	238	59.24	559	29.68	73.95	694	34.84
s382	270	60.00	828	41.21	75.93	1,187	49.92
s444	272	61.76	918	40.84	75.74	1,124	49.44
s386	310	74.19	1,570	62.34	81.29	1,844	73.63
s400	274	62.77	1,001	43.39	77.74	1,263	52.47
s420	254	62.99	896	83.80	79.92	1,035	78.98
s526	292	62.33	873	45.40	80.14	1,132	54.44
s526n	300	68.67	1,280	48.43	81.00	1,427	58.92
s1238	912	55.37	4,394	487.73	65.24	5,698	626.1
Total	4,974		17,825	1,094.94		21,998	1,339.05
Average		66.12			77.66		
Ratio					1.17	1.23	1.22

|*TW*|: Number of target wires.
%: Percentage of target wires having alternative wires.
|*AW*|: Number of alternative wires.
TIME: The CPU time, measured in seconds.

Comparison between ECR and the node substitution algorithms: Node merging and node addition and removal (NAR), were also made in Lam et al. (2012) despite the intrinsic differences between the two kinds of rewiring algorithms. Since NAR is designed for node but not wire substitution, the number of alternative wires for each node was counted as the number of the fanouts of the node times the number of node substitution solutions. The result is shown in Table 4.5.

4.2.6.2 Application on FPGA Technology Mapping

We showed in the previous section how the ECR approach is more efficient than the previous ATPG-based rewiring schemes. In this section, the results of applying different rewiring schemes to FPGA technology mapping are compared.

FPGA technology mapping is a process to map gates to lookup tables (LUTs)on an FPGA device. The basic programmable logic element in FPGA is the K-input LUT. It was proven in Tang et al. (2007) that the rewiring augmented technology mapping method, which is known as the incremental logic resynthesis (ILR) approach, could achieve further improvement on the level and area upon the outstanding technology mapping algorithm DAOMap (Chan and

Table 4.5 Comparison on combinational benchmarks between NAR and ECR

Benchmark	$\|TW\|$	NAR			ECR		
		%	$\|AW\|$	TIME	%	$\|AW\|$	% TIME
9sym-hdl	96	7.29	7	0.01	26.04	58	0.99
pcler8	126	0.00	0	0.03	28.57	106	4.1
f51m	190	6.84	39	0.07	49.47	441	8.5
comp	184	4.89	13	0.03	42.39	610	28.2
5xp1	178	5.06	24	0.07	50.56	452	6.86
b9_n2	166	0.60	2	0.01	46.99	240	5.24
my_adder	256	5.86	15	0.01	31.64	134	6.27
ttt2	274	2.55	21	0.07	43.80	521	13.85
term1	314	7.96	87	0.05	45.22	699	32.28
sao2-hdl	324	3.09	17	0.15	29.63	428	50.91
C432	320	6.56	23	0.07	40.94	545	43.97
duke2	566	1.77	54	0.87	44.35	2,230	144.44
alu2	662	4.23	132	1.52	47.89	2,749	276.02
C1908	706	0.71	7	0.23	22.95	291	148.37
misex3	840	5.83	220	2.46	41.31	3,170	365.32
C880	648	0.62	4	0.09	29.78	458	124.52
C1355	772	0.00	0	0.17	22.28	228	124.04
rot	942	0.53	15	0.15	42.25	1,046	227.36
x3	1,120	0.36	5	0.46	30.45	1,287	405.82
apex6	1,188	1.68	66	0.43	31.90	2,099	469.82
alu4	1,252	3.12	188	5.21	49.84	3,826	1,661.69
C2670	1,074	0.19	2	0.18	29.98	2,960	958.83
C3540	1,870	0.48	12	5.52	36.04	4,051	3,290.54
C5315	2,622	0.61	30	0.49	28.34	2,542	1,998.72
C7552	2,876	1.91	105	1.30	24.86	2,039	4,024.5
Total	19,566		1,088	19.63		33,210	14,421.16
Average		2.91			36.70		
Ratio					12.61	30.52	735.65

$|TW|$: Number of target wires.
%: Percentage of target wires having alternative wires.
$|AW|$: Number of alternative wires.
TIME: CPU time, in seconds.

Cong 2004). The logic-aware minimization technique ILR examines the set of available transformations and commits a transformation greedily if it can help to reduce the level or area of the circuit. In other words, this method explores all possible transformations to reduce the level and area until no more transformation is beneficial. As readers can imagine, the number of useful transformations depends on the number and quality of alternative wires. ILR is therefore very sensitive to the rewiring ability of the underlying rewiring engine.

In the following experiment, the main optimization objective is to reduce the level of the circuit, and the next important optimization objective is to minimize the area of the circuit. IRRA and ECR were used to power ILR, and each setup was evaluated individually. Table 4.6 gives the 4-LUT FPGA technology mapping results with regard to benchmarks that had already

Table 4.6 FPGA technology mapping for ILR combined with different rewiring engines on optimized benchmarks, LUT size = 4

Benchmark	DAOMap		ILR-DAOMap (IRRA)			ILR-DAOMap (ECR)		
	LV	|LUT|	LV	| LUT |	Time	LV	|LUT|	TIME
5xp1	4	38	3	19	327.92	3	19	255.57
9sym-hdl	4	14	4	10	44.61	3	10	33.07
alu2	11	144	8	109	5382.86	8	108	5889.61
b9_n2	3	43	3	42	145.55	3	42	155.59
C432	11	77	10	63	801.53	10	63	546.77
comp	5	41	5	30	247.77	5	29	311.87
my_adder	16	33	11	39	136.29	11	39	113.06
sao2-hdl	11	84	7	39	687.54	7	37	430.34
term1	5	64	4	28	655.25	4	28	474.9
ttt2	4	54	4	40	463.33	3	49	379.63
Total	74	592	59	419	8892.65	57	424	8590.41
Ratio1	1		79.7%			77.0%		
Ratio2		1		70.8%			71.6%	
Ratio3					1			96.6%

LV: Level of corresponding benchmark.
|LUT|: Number of LUTs.
TIME: CPU time, in seconds.

been optimized by rewriting. It was found that almost half of the benchmarks could be further optimized by IRRA and ECR. The levels of the two benchmarks *9sym-hdl* and *ttt2* achieved by ECR was smaller than those achieved by IRRA. Area reduction was observed in three of the eight remaining benchmarks. On average, the level of circuit was reduced by 20.3% (1–79.7%) by IRRA while 23%(1–77.0%) by ECR. Both these two methods could achieve 30% more LUT reduction than when applying rewriting alone.

The experiments demonstrate that ECR could overcome the limitation of the conventional redundancy addition and removal. The ECR approach is unique in that dynamic dominators are used to locate more alternative wires with high efficiency. The notion of error cancellation is explored and efficiently applied in this approach to expand the scope of locating alternative wires.

4.3 FECR

Error-cancellation-based rewiring techniques were further improved in later researches. The authors of flow-graph-based error cancellation rewiring (FECR) (Yang et al. 2012) analyzed the theory on error cancellation and proposed a more general and systematic rewiring approach. FECR is a general 1-to-k (using k alternative wires to replace 1 target wire) rewiring scheme where a flow graph is constructed to model error propagations such that the error caused by the target wire removal is cancelled at the minimum cut(s) of the graph. To cancel the error, a rectification network is temporarily added into the circuit. A NAR-based algorithm (Chen and Wang 2010a) is then employed to find the source nodes, which can be existing nodes in the circuit or new nodes, to replace the rectification network.

Although the approach of ECR (Yang et al. 2010) is already more general than the traditional RAR schemes, it is still limited to the "one alternative wire for one target wire" constraint. An FECR algorithm was proposed in Yang et al. (2012) to relax the constraint and to explore a much broader solution space. The authors were inspired by the ideas introduced in global flow optimization (GFO) and implication flow graph (IFG) (Wu and Chang 2006), in which wire reconnections are modeled as a flow graph optimization problem to be solved by "maxflow-mincut" algorithms.

In FECR, the error propagation paths are first identified and modeled as a graph. Rectification networks are then added into the circuit to cancel the errors along the propagating paths. The authors of this new rewiring scheme suggested the determination of the min-cuts of the error propagation graph and adoption of the corresponding nodes in the circuit as the destination nodes of the rectification networks. The flow will be introduced in more detail with illustrative examples in the following text.

4.3.1 Error Flow Graph Construction

After collecting the mandatory assignments for activating and propagating the error $\epsilon_r.w_t$, the *error flow graph* (also called *error propagation graph*) is constructed as follows:

1. The target wire's sink node is added as the source node (root) s of the error flow graph. This node is given infinite (inf) weight.
2. All nodes having implied values D or $\bar{D}$ are added. Initially, all such nodes have weights $= 1$.
3. A sink node t with infinite weight is added to the graph.
4. Edges are inserted between nodes following the error propagation paths of the circuit. If a fanout of a node has no MA, the node is connected with the sink directly.

For example, suppose that $b \to g1$ is the target wire in the circuit in Figure 4.19. To activate the w_t_error and propagate it through $g1$, the logic values at wire $b \to g1$ and a are set to $1/0$ and 0, respectively. After implications, the following MAs are obtained: $\{a = 0, b = 1/0, g1 = 1/0, g3 = 1, g4 = 0, g12 = 0, g5 = 1/0, g7 = 1/0, g8 = 1/0\}$. The FMAs

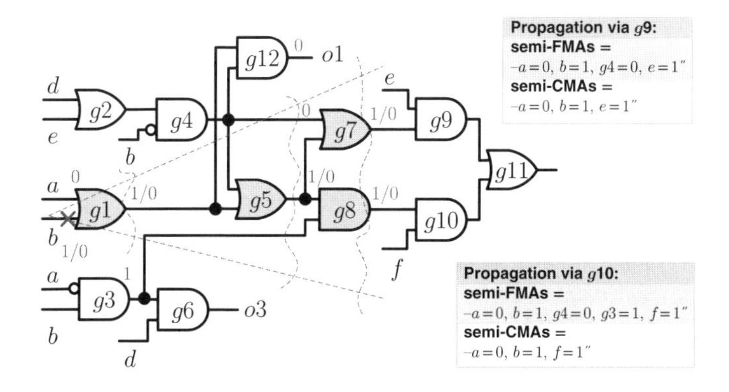

Figure 4.19 Error propagation of w_t_error

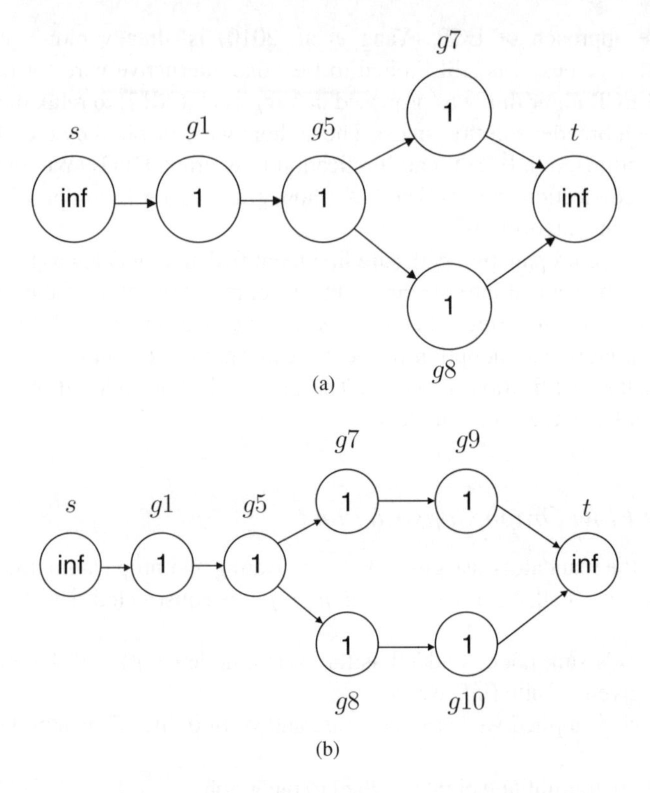

Figure 4.20 Error flow graphs (a) without semi-MA and (b) with semi-MA

are $\{a = 0, b = 1, g4 = 0\}$ and the CMAs are $\{a = 0, b = 1\}$. Figure 4.20(a) illustrates the error flow graph created according to the construction steps. Although $g12$ is in the transitive fanout cone of $\epsilon_r(w_t)$ and has an MA, it does not exist in the error flow graph because the value at $g12$ remains 0 in both the good and the faulty circuits. This means the error cannot be propagated through $g12$, and therefore it should be excluded from the error flow graph.

Min-cuts of sizes 1 and 2 will then be determined from the error flow graph to identify candidate destination nodes of alternative wires.

4.3.2 Destination Node Identification

Definition 4.8 *Given a network, a node set S is an E-frontier of a wire n's stuck-at-1 or stuck-at-0 fault if:*

1. *Every node in S must have an MA.*
2. *For each path from node s_i ($s_i \in S$) to any primary output o: $s_i \to n_1 \to n_2 \to \cdots \to n_k \to o$, any node n_j ($j \in [1, k]$) is not in S.*
3. *Every error propagation path from wire n to any primary output must go through at least one node in S.*
4. *Each node in S is reachable from n.*

Definition 4.9 *Given a network C, a stuck-at fault s, and a node t which is in the transitive fanout cone of s, a node n has a semi-MA x (x is 0 or 1) with respect to t if:*

1. *V is the set of test vectors that can activate s and propagate it to the primary outputs via node t, and*
2. *node n has the same value x in all test vectors in V.*

For Definition 4.9, semi-FMA and semi-CMA are defined accordingly. The logical conjunction of semi-CMAs with respect to a node t, AND(semi-CMA), represents the test vectors that can propagate the error via t.

According to Definition 4.8, each min-cut in the error flow graph is an E-frontier. Then, an error will not be observable if it can be stopped from propagating through an E-frontier. All E-frontiers of various sizes can be determined by maxflow-mincut algorithms. With reference to the example in Figure 4.19, the E-frontiers are found to be $\{g1\}$ and $\{g5\}$ when the cut size is 1. $\{g7, g8\}$ is an E-frontier when cut size is 2. The error can be cancelled either at $\{g1\}$, or $\{g5\}$, or $\{g7, g8\}$. When the number of nodes in an E-frontier is more than 1, multiple errors may need to be injected into the circuit to cancel the error $\epsilon_r(w_t)$. As illustrated in Figure 4.21(b), three errors $e1$, $e2$, and $e3$ are required to cancel one another in their common fanout cone.

Regarding the example, the FMAs are $\{a = 0, b = 1, g4 = 0\}$. Assume that the E-frontier is $\{g7, g8\}$. If the error $\epsilon_r(w_t)$ is to be propagated via $g9$, e has to be assigned with 1. Semi-FMAs are found to be $\{a = 0, b = 1, g4 = 0, e = 1\}$. The corresponding semi-CMAs are then $\{a = 0, b = 1, e = 1\}$. On the other hand, if the error $\epsilon_r(w_t)$ is to be propagated via node $g10$, the node f has to be set to 1. The semi-FMAs are then $\{a = 0, b = 1, g4 = 0, g3 = 1, f = 1\}$, and the corresponding semi-CMAs are $\{a = 0, b = 1, f = 1\}$. The error flow graph is illustrated in Figure 4.20(b).

After finding the E-frontiers, semi-CMAs, and semi-FMAs, rectification networks which are constructed as AND(semi-CMA)s can be added at E-frontiers or nodes with semi-FMAs to cancel any error. Candidate alternative wire destinations can be:

1. nodes in the E-frontiers
2. nodes that have been assigned semi-FMAs.

If there is only one node in the E-frontier and that node is the alternative wire's destination, the destination node is valid. Otherwise, additional consistency check must be done to ensure the validity of the rewired circuit. Theorems related to destination node identification are discussed below.

Theorem 4.6 *If a node m has semi-FMAs in different propagation paths via the nodes in the same E-frontier, m is not the only candidate destination (more than one candidate destinations are necessary) for cancelling the error if the following conditions are satisfied:*

1. *m has different semi-FMA values for different propagation paths;*
2. *The intersection Γ of the test vectors corresponding to the different propagation paths is nonempty.*

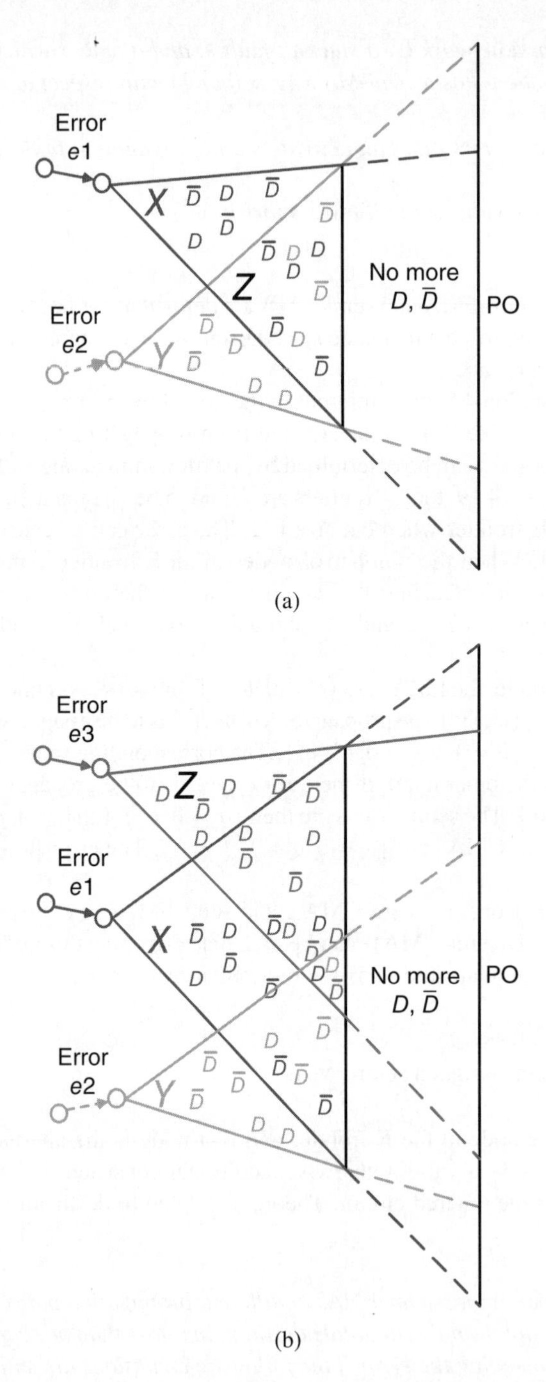

Figure 4.21 Structural view of error cancellation. (a) 1-to-1 error cancellation; (b) multi-error cancellation

Proof. Suppose the semi-FMA for m is 0 when the corresponding E-frontier node is v. And in another propagation path, when the corresponding E-frontier node is u, m's semi-FMA is 1. Let $\Gamma 1$ denote the test vectors to propagate $\epsilon_r(w_t)$ through node v. Also, let $\Gamma 2$ denote the test vectors to propagate $\epsilon_r(w_t)$ through node u. Then, Γ in the second condition is $\Gamma = \Gamma 1 \cap \Gamma 2$. Let $t \in \Gamma$.

When t is applied to test $\epsilon_r(w_t)$, if rectification is made at node m only, the error cannot be cancelled no matter what the value of m. This is because when the value of m is 0, error $\epsilon_r(w_t)$ can always be propagated through the node v or otherwise through the node u. ■

More than one candidate destination is required to correct $\epsilon_r(w_t)$ even if the E-frontier has only one node. Theorem 4.6 gives the conditions for filtering out impossible single candidate alternative wire destinations. For instance, if the E-frontier is $\{g7, g8\}$ in Figure 4.19, each of $g7$ and $g8$ cannot be the only single candidate alternative wire destination for rectifying the $\epsilon_r(w_t)$.

Algorithm 4.8: Flow-graph-based error cancellation rewiring (FECR)

```
    input : circuit C, target wire w_t
    output: rewiring solution RS
 1  begin
 2  |  MA = stuckAtFaultTest(w_t);
 3  |  efg = constructErrorFlowGraph(MA);
 4  |  S = findEFrontierSets(efg);
 5  |  foreach E-frontier S_i in S do
 6  |  |  foreach E-frontier node S_ij in S_i do
 7  |  |  |  semiFMA = findForcedSemiMAs(S_ij);
 8  |  |  |  insert S_ij, semiFMA into CD_ij;
 9  |  |  |  foreach candidate destination node C_d in CD_ij do
10  |  |  |  |  if singleDestinationValid(w_t, S_i, C_d) then
11  |  |  |  |  |  insert C_d into SD_ig;
12  |  |  |  insert SD_ij into SD_i;
13  |  |  D_i = validateConsistentDestinationGroupSet(SD_i);
14  |  |  foreach consistent destination group D_ij in D_i do
15  |  |  |  SRS = identifySourceNode(D_ij);
16  |  |  |  insert SRS into RS;
17  |  return RS;
18  end
```

Theorem 4.7 *Suppose there are two nodes u and v in the same E-frontier. The corresponding semi-MAs for each of these E-frontier nodes are found by propagating the error $\epsilon_r(w_t)$ through u and v, respectively. Denote the AND(semi-CMA) of a node n as AND(semi-CMA(n)). If*

$\epsilon_r(w_t)$ is to be corrected at both nodes u and v at the same time, the necessity of consistency check is based on the following conditions:

1. *If AND(semi-CMA(u)) $\cap$ AND(semi-CMA(v)) $= \Phi$, consistency check can be exempted.*
2. *If AND(semi-CMA(u)) $\cap$ AND(semi-CMA(v)) $\neq \Phi$, consistency check is necessary for AND(semi-CMA(u)) $\cap$ AND(semi-CMA(v)).*

Proof. Suppose that the intersection is empty. Since AND(semi-CMA(u)) represents the necessary condition for all the test vectors that can distinguish between the good circuit and rewired circuit when the $\epsilon_r(w_t)$ is propagated through u, it is impossible for AND(semi-CMA(v)) to affect the error propagation through u. The two test vectors cannot be effective simultaneously and therefore a consistency check is not necessary. ■

For instance, a consistency check is necessary if the w_t_error is rectified at both $\{g7, g8\}$ in the example shown in Figure 4.19. The FECR algorithm is summarized in Algorithm 4.8.

4.3.3 Source Node Identification

After identifying candidate destination nodes, the next task is to figure out how the rectification networks should be constructed.

4.3.3.1 First Approach

For all the candidate destination nodes, a common rectification network can be constructed using the AND(CMA) obtained from the stuck-at fault test of the error $\epsilon_r(w_t)$. Then, substitution nodes for the rectification network are identified as the alternative wires' sources using the techniques presented in Chen and Wang (2009, 2010a).

Referring to the example in Figures 4.22 and 4.23, if $\{g7, g8\}$ are the candidate destination nodes, the circuit after adding the rectification network is shown in Figure 4.22(a). The substitution node for the rectification network is found to be $g3$. The target wire $b \rightarrow g1$ is removed. Figure 4.22(b) shows the rewired circuit. The circuit can be further simplified. For example, the gated $g8$ is obviously redundant after the addition of the rectification network.

4.3.3.2 Second Approach

Since it is not limited to have only one candidate destination node in FECR, the rectification condition in FECR is not as restrictive as that in the ECR approach. It may be advantageous to relax the AND(CMA) rectification scheme.

AND(semi-CMA) is a subset of AND(CMA). From the example in Figure 4.19, it is known that the CMAs are $\{a = 0, b = 1\}$ in the ECR approach. The rectification network built upon the CMAs and the corresponding single alternative wire are valid only if none of the test vectors in $\{a = 0, b = 1\}$ can distinguish between the good circuit and the rewired circuit. Unlike ECR, the test vectors in FECR are divided into three parts: (a) $\{a = 0, b = 1, e = 1\}$

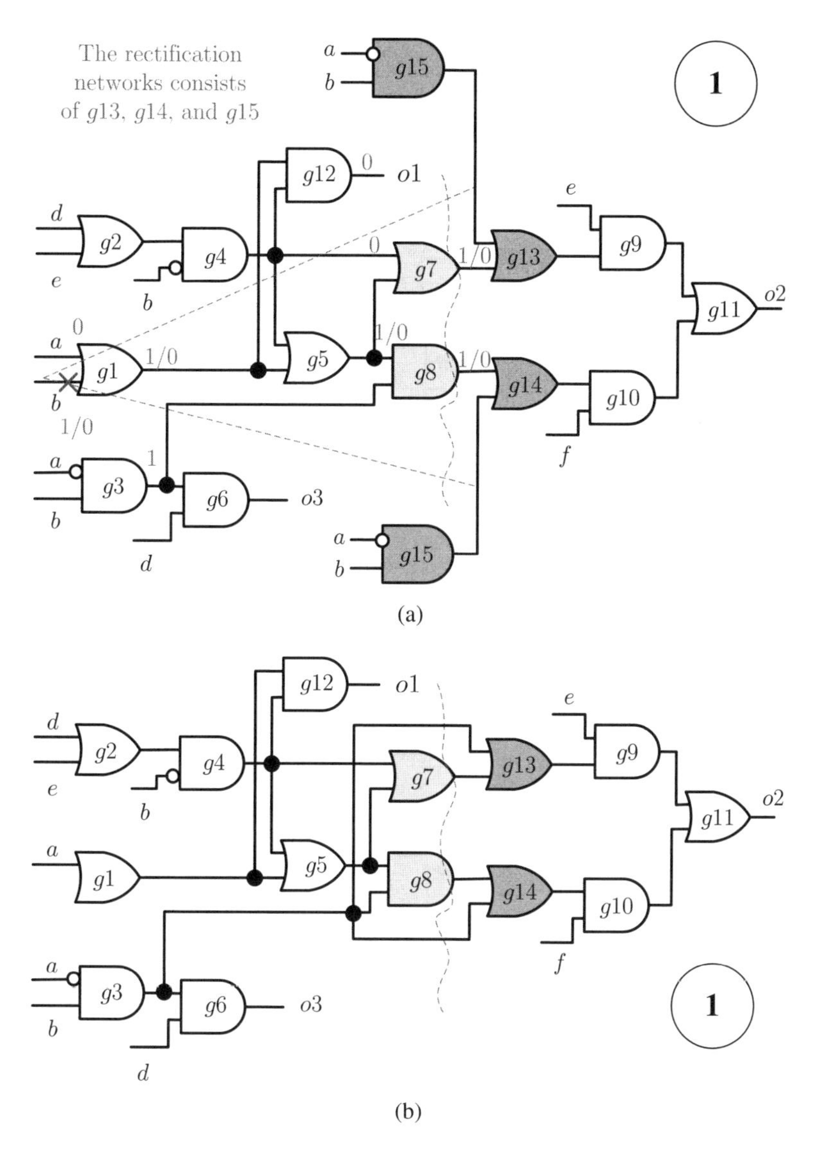

Figure 4.22 Example of flow-graph-based error cancellation rewiring (FECR). (a) Addition of the rectification network –first approach; (b) rewired circuit –first approach

and (b) $\{a = 0, b = 1, f = 1\}$ and (c) the don't cares: $\{a = 0, b = 1, e = f = 0\}$. The rewired circuit is functionally equivalent to the original circuit only if none of the test vectors in part (a) and part (b) can distinguish between these two circuits.

Rectification networks for each of the candidate destination nodes can be therefore constructed using the corresponding AND(semi-CMA) instead. Figure 4.23(a) depicts this approach and shows the rewired circuit after rectification network substitutions.

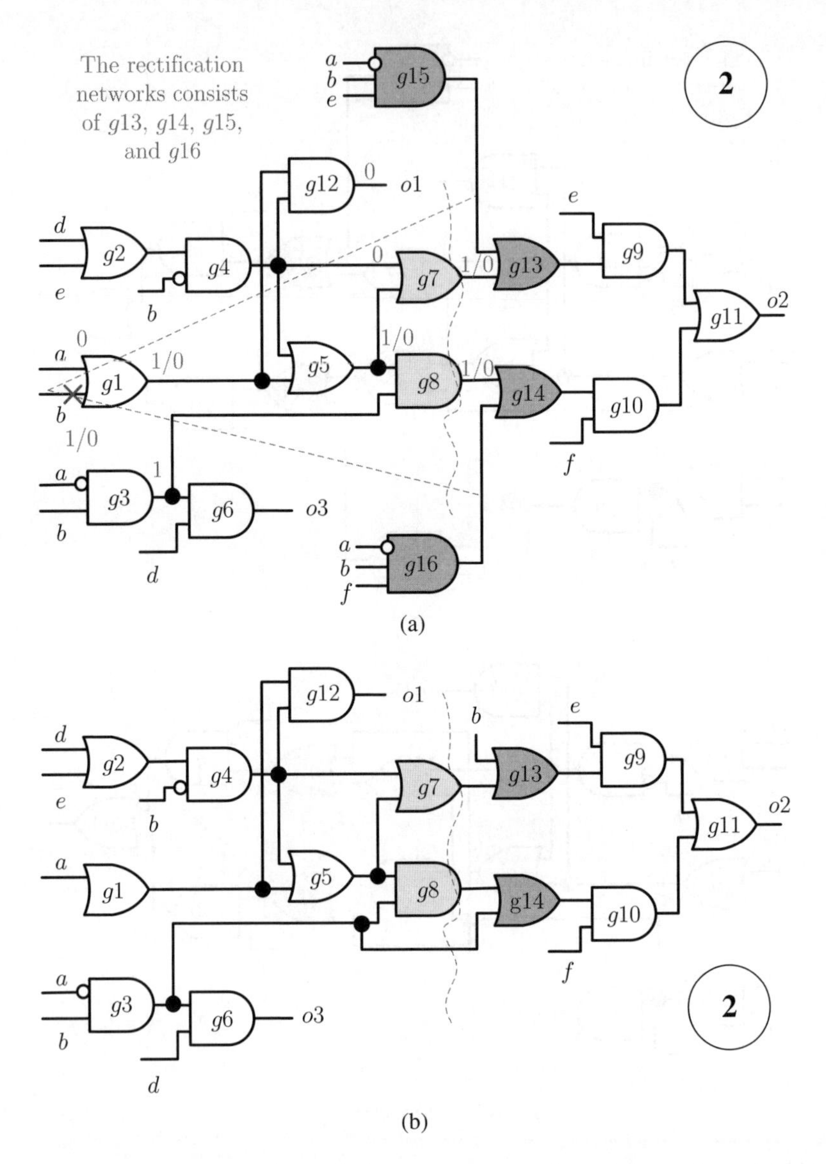

Figure 4.23 Example of flow-graph-based error cancellation rewiring (FECR). (a) Addition of the rectification network –second approach; (b) rewired circuit –second approach

4.3.4 ECR is a Special Case of FECR

Referring to Definition 4.8, an E-frontier is a set of nodes at which the error $\epsilon_r(w_t)$ can be cancelled. A single node in an E-frontier is essentially a testing dominator. Therefore, the circuit transformations found by ECR is a subset of FECR. Since RAR is a special case of ECR, both RAR and ECR are subsets of FECR.

The structural views on 1-to-1 and 1-to-k rewirings are shown in Figure 4.21. As depicted in Figure 4.21(a), the error effects of e_1 and e_2 cancel each other mutually in the region Z. No error effects can be observed at any primary output. As illustrated in Figure 4.21(b), the errors caused by target wire removal can be cancelled by adding multiple errors into the circuit.

4.3.5 Complexity Analysis of FECR

Assume that the number of nodes in a circuit is n and the runtime for a stuck-at fault test is normally no more than a constant T (in practice). Suppose a circuit is decomposed into two-input gates. Then, the number of target wires is $2n$. The number of nodes in the error propagation graph is bounded by n. Further, suppose that only min-cuts having no more than two nodes are determined for efficiency. The time complexity for finding a min-cut is $O(n^2)$. The total time consumed for finding min-cuts is $\mid$ E-frontier $\mid \cdot O(n^2)$. However, the number of nodes in the error propagation graph is not related to n in most cases. $\mid$ E-frontier $\mid \cdot O(n^2)$ is the upper bound for finding min-cuts. For each candidate destination, the time consumed on the function `singleDestinationValid` is $O(n)$. For each pair of candidate destinations, the time consumed on `validateConsistentDestinationGroupSet` is also $O(n)$. Let $\mid CD \mid$ denote the number of candidate destinations for each E-frontier node. The total time consumed for finding valid pairs of destinations is $\mid$ E-frontier $\mid \cdot \mid CD \mid^2 \cdot O(n)$. For each destination pair, the time consumed on finding the source nodes is $\mid source\ nodes \mid \cdot \mid MA \mid \cdot T$. The total time consumed on finding the source nodes is then $\mid source\ nodes \mid \cdot \mid MA \mid \cdot T \cdot \mid destination\ pairs \mid \cdot \mid$ E-frontier $\mid$. The time complexity of the FECR algorithm is thus $\mid$ E-frontier $\mid \cdot O(n^2) + \mid$ E-frontier $\mid \cdot \mid CD \mid^2 \cdot O(n) + \mid source\ nodes \mid \cdot \mid MA \mid \cdot T \cdot \mid destination\ pairs \mid \cdot \mid$ E-frontier $\mid$.

Theoretically, the number of MAs for a stuck-at fault test could be $O(n)$. $\mid$ E-frontier $\mid \cdot \mid CD \mid^2$, $\mid destination\ \ pairs \mid \cdot \mid$ E-frontier $\mid$, and $\mid source\ \ nodes \mid$ could be $O(n^2)$ in the worst case. Thus the complexity is bounded by $O(n^5)$. However, in practice, most of these items are hardly related to n. Empirical analysis has shown that the actual cost of the whole process is very close to $O(n)$.

4.3.6 Experimental Result

Table 4.7 shows the result of the comparison between ECR and FECR. The algorithm was implemented in C++ and tested with ISCAS and MCNC benchmarks that had been optimized by ABC (Berkeley Logic Synthesis and Verification Group). Experiments were performed on a 2.8-GHz machine with 1 GB RAM.

In Table 4.7, the column TW represents the number of target wires in each circuit. The third column gives the percentage of target wires having single alternative wires in ECR. The fourth column lists the number of alternative wires found by ECR. The fifth column presents the CPU time measured in seconds. The percentage of target wires having alternative wires in FECR is listed in the sixth column. The seventh column gives the number of single alternative wires found by FECR. The eighth column gives the number of alternative wires with single destinations. The ninth column shows the number of alternative wires with pair destinations. The last column lists the CPU time.

Regarding the percentage of target wires whose alternative wires could be identified, around 35% was reported in Chang and Marek-Sadowska (2001) and over 60% was reported in Lin and Wang (2009) for circuits that had not been optimized. For circuits optimized by ABC

Table 4.7 Comparison between ECR and FECR

Benchmark	$\vert TW\vert$	ECR			FECR				
		%	$\vert AW\vert$	Time (s)	%	$\vert AW$ (1-to-1)$\vert$	Single destination	Pair destination	Time (s)
9sym-hdl	96	26.04	58	0.99	81.25	55	989	2,962	7.03
pcler8	126	28.57	106	4.1	80.95	150	2,017	0	8.68
f51m	190	49.47	441	8.5	98.95	445	20,589	334,995	183
comp	184	42.39	610	28.2	74.46	334	1,835	8,875	187.56
5xp1	178	50.56	452	6.86	98.88	460	21,141	269,695	140.88
b9_n2	166	46.99	240	5.24	94.58	306	2,327	3,097	27.25
my_adder	256	31.64	134	6.27	86.72	136	2,669	2,677	26.88
ttt2	274	43.80	521	13.85	95.26	548	18,064	47,508	74.72
term1	314	45.22	699	32.28	95.54	793	19,399	179,782	389.07
sao2-hdl	324	29.63	428	50.91	92.90	343	34,558	392,812	781.97
C432	320	40.94	545	43.97	91.25	732	7,390	1,841	95.1
C1908	706	22.95	291	148.37	97.03	302	6,677	11,156	490.3
C880	648	29.78	458	124.52	94.14	551	7,067	2,598	259.27
C1355	772	22.28	228	124.04	98.45	260	4,506	8,002	423.13
rot	942	42.25	1,046	227.36	93.63	1,146	13,342	19,948	745.94
x3	1,120	30.45	1,287	405.82	97.23	1,247	27,766	57,594	1,084.97
apex6	1,188	31.90	2,099	469.82	97.47	2,196	41,408	112,536	1,984.01
Average		36.17			92.27				
Ratio		1			2.49				

$|TW|$: Number of target wires in the circuit.
% (for ECR): Percentage of target wires which have alternative wire.
$|AW|$ (for ECR): Number of total alternative wires found.
Time (s): CPU time, measured in seconds.
% (for FECR): Percentage of target wires that are removable.
$|AW(\text{1-to-1})|$ (for FECR): Number of all single alternative wires.
Single destination: Number of solutions with single destination.
Pair destination: Number of solutions with pair destination.

(Berkeley Logic Synthesis and Verification Group) (to reduce structural bias and literals), the percentage was found to be around 20% for Lin and Wang (2009), 36.17% for ECR, and 92.27% for FECR. The number of 1-to-1 alternative wires found by FECR matched with that found by ECR generally. For most benchmarks, the CPU time taken by FECR was several to a dozen times larger than that taken by ECR. The CPU time for some circuits was particularly high because exhaustive search was applied to find as many alternative solutions as possible and to reflect the potentially very large solution space. In practice, an upper bound can be imposed on the number of alternative wires to be identified.

The authors of FECR (Yang et al. 2012) have shown that virtually every single wire in a circuit is removable at the maximum cost of several alternative wires and that the alternative wires can be found within a reasonable amount of time. This significant progress allows rewiring to be a much more flexible logic synthesis technique that provides today's various challenging EDA problems one extra last resort to explore.

4.4 Cut-Based Error Cancellation Rewiring

In this chapter, the theory on error cancellation is further explored. A systematic cut-based error cancellation rewiring (CECR) (Wei et al. 2013) solution is discussed. The objective of CECR is to gain a *rewiring rate* (percentage of target wires that are removable) comparable to the most recent rewiring scheme FECR while using a much smaller CPU time (32× speedup) and make the scheme scalable for realistic and large industrial circuits which may consist of millions of gates.

In the CECR scheme, the error propagation algorithm is extended to find more mandatory assignments as well as rewiring solutions. An effective cut enumeration algorithm is used in the scheme to reduce the complexity significantly. Finally, it makes use of a windowing approach to make the scheme scalable for large circuits. The impact of the windowing approach on the rewiring performance is also thoroughly explored.

4.4.1 Preliminaries

Previous ATPG-based rewiring schemes rely on dominators to find MAs and efficiently identify alternative wires. If a gate is a dominator of the target wire, it can be considered as the destination node of a candidate alternative wire. The source of the alternative wire is another gate with MA.

However, the traditional structural dominator information does not provide sufficient information to determine more alternative wires. In Krieger et al. (1991), *dynamic dominator* was defined. Since structural dominators are just special cases of dynamic dominators, some even nonredundant alternative wires can be found if dynamic dominators are allowed to serve as the potential alternative wires' sinks (Yang et al. 2010). The use of dynamic dominators as alternative wires' sinks makes ECR superior to RAR-based rewiring schemes in rewiring ability. In FECR, a one-to-many rewiring scheme was introduced. It allows more alternative wires to be found and more target wires to be substituted.

The CECR rewiring framework is defined using the definitions below as a foundation.

Definition 4.10 *Given a Boolean network, a wire $w_t(u \rightarrow v)$, and its removal error $\epsilon_r(w_t)$, the blocking cut (B-cut) of the wire is defined to be the blocking cut of node v. Its error-cut (E-cut) is a set of nodes in the B-cut whose mandatory assignments are neither* 1 *nor* 0. *In other words, the nodes of an E-cut allow the wire removal error $\epsilon_r(w_t)$ to propagate through them.*

The definitions are illustrated in Figure 4.24. Assume that an error w_t_error is propagated through some gates in $E\text{-}cut1$. In Figure 4.24(a), the nodes labeled with the letter **E**, which form an $E\text{-}cut$, are gates through which the error propagates. They also form a $B\text{-}cut$ together with other nodes labeled with the letter **C**. Another example is shown in Figure 4.24(b) where a blocking cut $B\text{-}cut2$ contains an $E\text{-}cut2$.

Figure 4.25 gives an example to show the relationship between dominators, dynamic dominators, and error cuts. The MA values of the inputs and outputs of all gates are also indicated in the figure. The error cut $E\text{-}cut1$ is equal to a dominator because $|B\text{-}cut1| = |E\text{-}cut1| = 1$. The error cut $E\text{-}cut3$ is equal to a dynamic dominator because $|B\text{-}cut3| > |E\text{-}cut3| = 1$.

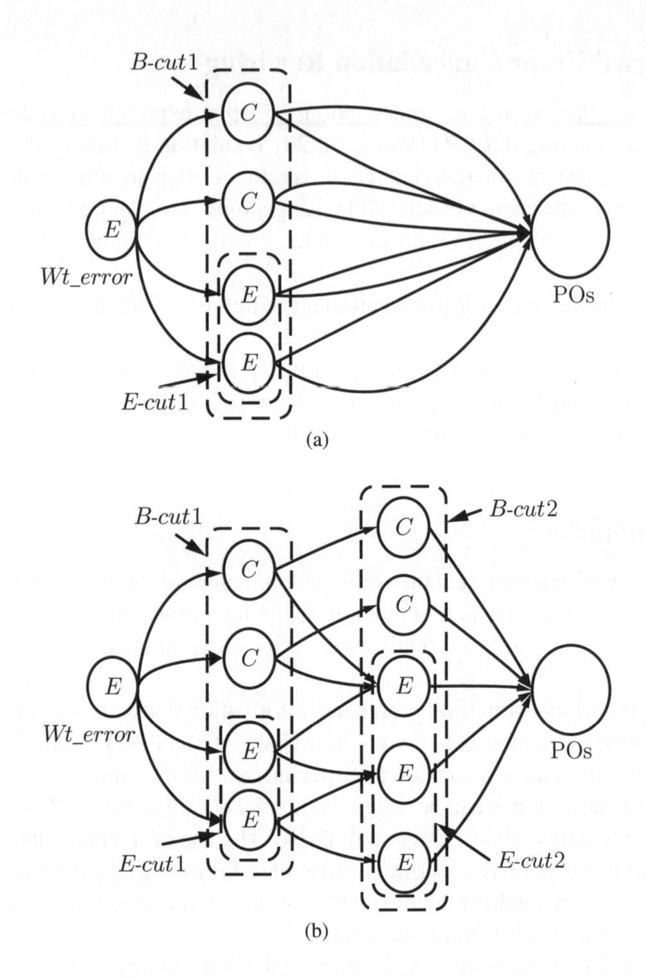

Figure 4.24 Example of $B\text{-}cut$ and $E\text{-}cut$

The FECR scheme also adopts the concept of cuts, where the propagation of an error is modeled as a graph. The min-cuts of the error flow graph are identified and serve as the set of destination nodes of alternative wires. They are essentially error cuts in the context of CECR. In CECR, the use of $B\text{-}cut$ and $E\text{-}cut$ to determine the destinations nodes of alternative wires with an efficient algorithm was analyzed.

Suppose there is a wire $w_t(u \to v)$ and its removal error $\epsilon_r(w_t)$ is a Boolean network. Clearly, this error must pass through at least one gate in any of its $E\text{-}cut$ if the error is testable (can be observed at some primary outputs). This implies that a target wire can be removed without changing the functionality of the circuit if and only if all errors can be blocked (cancelled) at an error cut. And this also implies that identifying more error cuts can help to identify more alternative wires. The CECR scheme was developed based on the idea of cancelling errors at error cuts. A major challenge in this scheme is to find the error cuts efficiently.

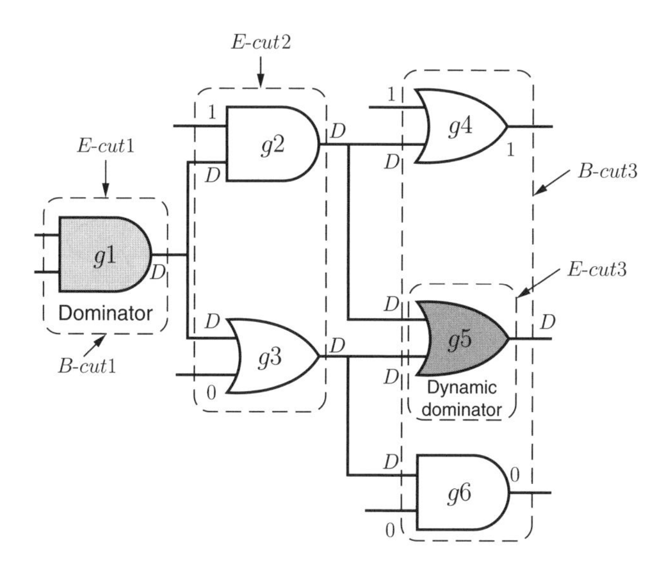

Figure 4.25 Relationship between dominators and error cuts

4.4.2 Error Frontier

In this section, a new error propagation scheme to identify MAs is presented. The dominator-based error propagation scheme adopted in previous ATPG-based rewiring schemes is first summarized for comparison. This is then followed by the new error propagation scheme adopted in CECR.

4.4.2.1 Dominator-Based Error Propagation

In previous ATPG-based rewiring schemes, dominators were considered one of the most crucial and useful structural information. An example of dominators and MAs is shown in Figure 4.26. The original circuit is shown in Figure 4.26(a) where the target wire is $w_t(b \rightarrow g1)$. All dominators ($g1$ and $g9$) of the wire removal error $c_r(w_t)$ are located, and the values of side inputs of the dominators are assigned as shown in Figure 4.26(b). Then, all MAs identified are shown in Figure 4.26(c).

4.4.2.2 Error-Frontier-Based Error Propagation

In Figure 4.26(c), the error D is propagated from $b \rightarrow g1$ to primary outputs. In order to achieve this, the primary input d must be assigned with the value 1 so that the error value propagated from gate $g7$ can go through gate $g10$ and finally to the primary output x. Since gate $g10$ is not a dominator, the error propagation described above cannot occur in dominator-based error propagation. In CECR, a new error propagation scheme that can handle this kind of situation was proposed. An *error frontier* (E-frontier) is defined as follows:

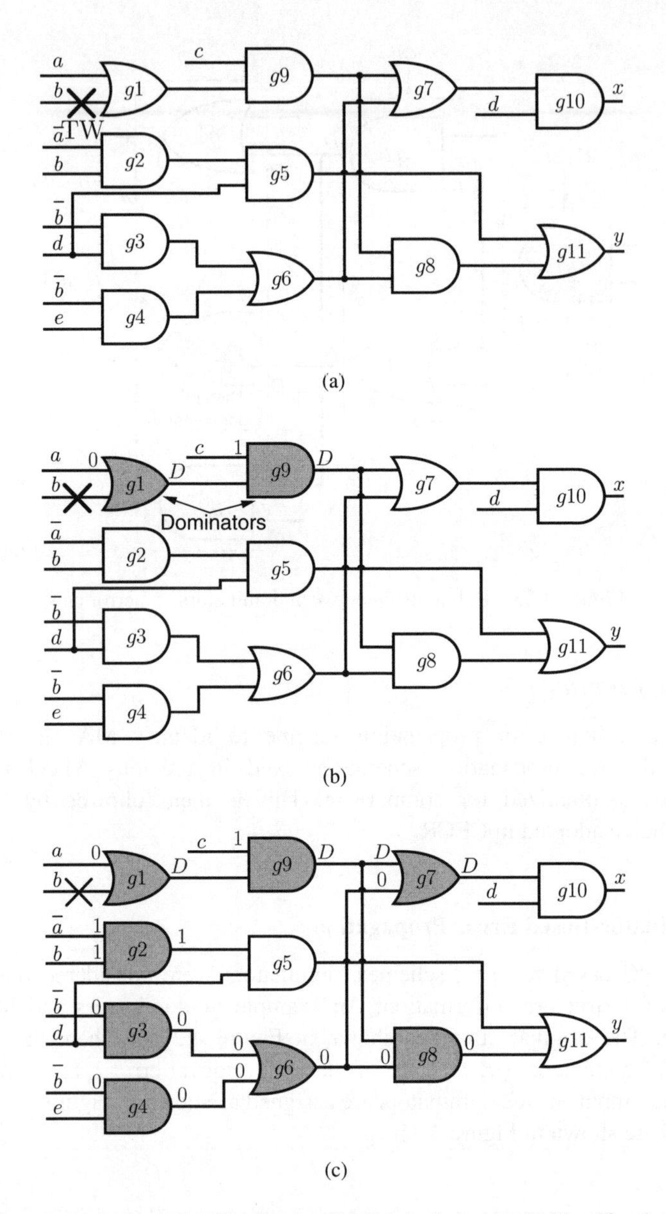

Figure 4.26 Error propagation using dominators. (a) Original circuit; (b) identification of dominators; (c) identification of mandatory assignments

Definition 4.11 *Given a Boolean network, a wire w_t, and its removal error $\epsilon_r(w_t)$, an error cut is an error frontier $E\text{-}frontier$ if every gate in this error cut has an error-MA (D or $\bar{D}$) and is either connected to a primary output or a node with an undetermined implication value (X).*

An error frontier of an error $\epsilon_r(w_t)$ indicates the current boundary beyond which the error propagation cannot proceed. The authors of CECR suggested a way to "shift" error frontiers

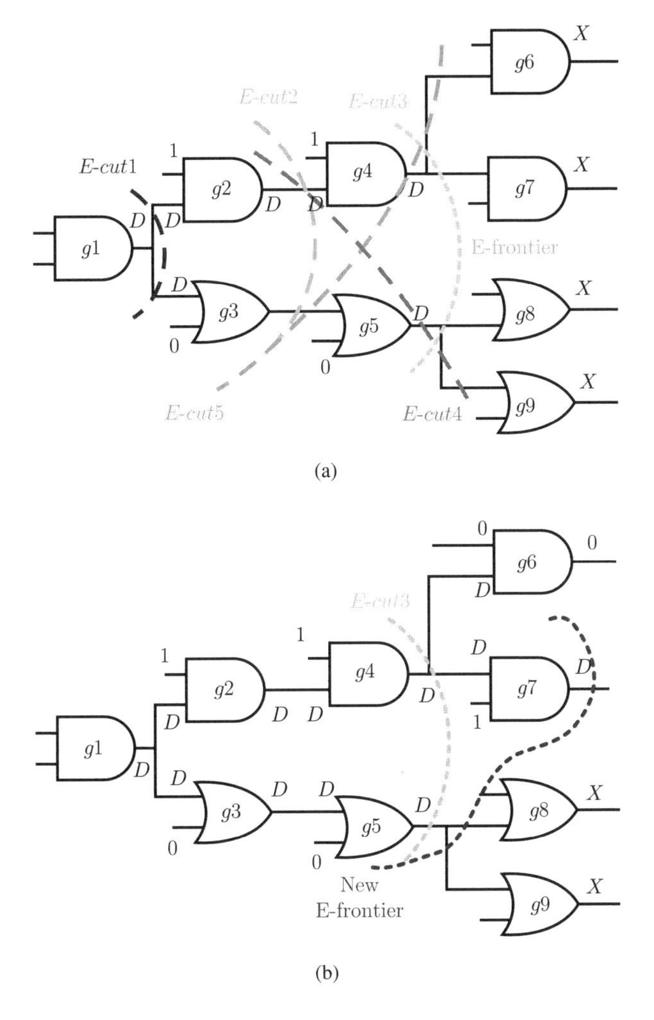

Figure 4.27 Example of $E\text{-}cuts$ and E-frontier shifting

toward the primary outputs by identifying more mandatory assignments. For example, given the known MAs shown in Figure 4.27(a), we can locate five different error-cuts. Suppose the current E-frontier is $E\text{-}cut3$. Later, when more MAs have been found (as in Figure 4.27(b)), the new E-frontier can be shifted to $\{g5, g7\}$.

A more complex example is shown in Figure 4.28, which illustrates the flow of error frontier shifting. In Figure 4.28(a), after an error is injected into the circuit, the initial error frontier is $E\text{-}frontier1$, which contains $g1$ only. Then, after more MAs have been assigned to activate and propagate the error, the E-frontier is shifted to $E\text{-}frontier2$ and $E\text{-}frontier3$ (Figure 4.28(b)). Figure 4.28(c) shows the final result of the error propagation. Note that the MAs of $g5$, $g10$, and $g11$ assigned by the CECR procedure cannot be obtained by the dominator-based propagation as shown in Figure 4.28(c).

The whole flow of error-frontier-based error propagation procedure is described in Figure 4.29.

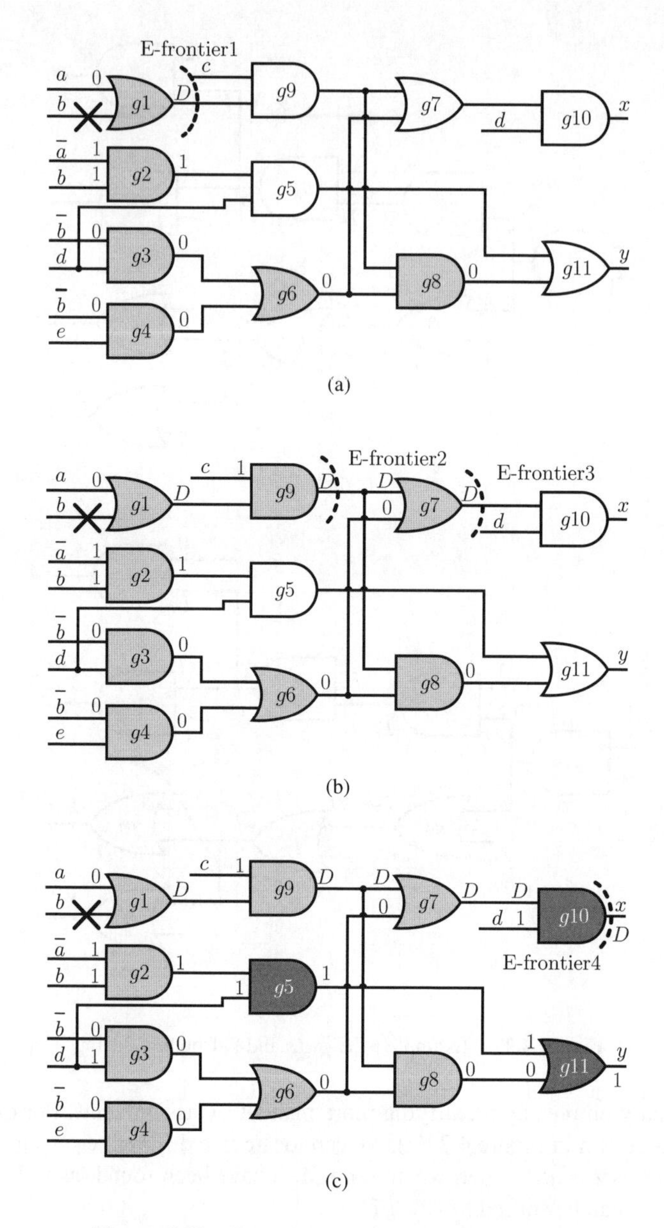

Figure 4.28 Error propagation using error frontier

4.4.2.3 Error Frontier Shifting

In this section, the different algorithms for shifting an E-frontier node to a single new E-frontier node (Figure 4.30) or multiple new E-frontier nodes (Figure 4.31) are explained. An E-frontier containing only one gate is called "a single-node frontier"; otherwise it is a "multi-node frontier."

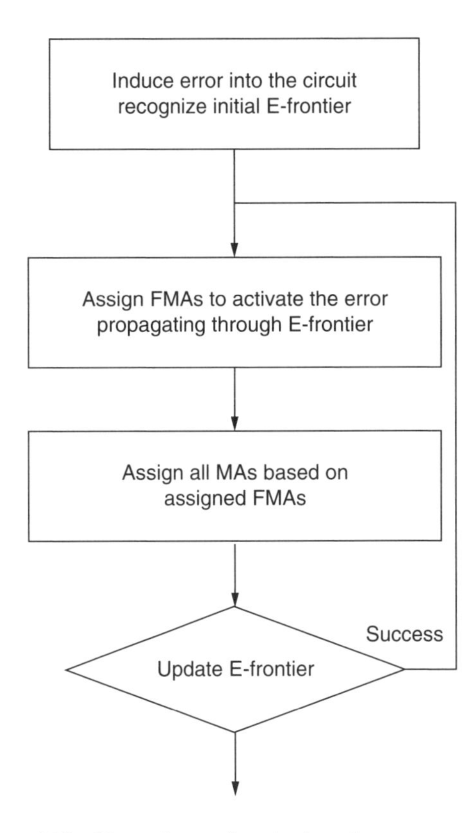

Figure 4.29 Flow of error-frontier-based error propagation

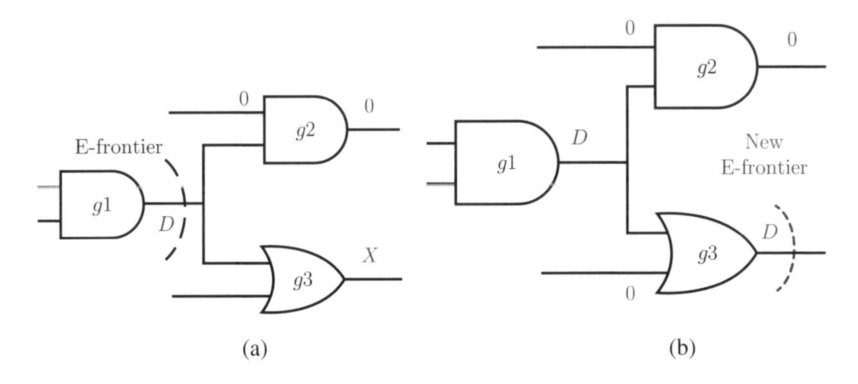

Figure 4.30 Locating new single-node frontier. (a) Original error frontier; (b) new error frontier

In Figure 4.30(a), the current E-frontier node has two fanout gates $g2$, $g3$. Since the upper fanout gate $g2$ has already got an MA value 0, it is impossible to propagate the error-MA (D or $\bar{D}$) further through it. Hence, the current E-frontier node can only be shifted to its lower fanout node $g3$ by assigning the side input of $g3$ to the noncontrolling value 0, which is depicted in Figure 4.30(b).

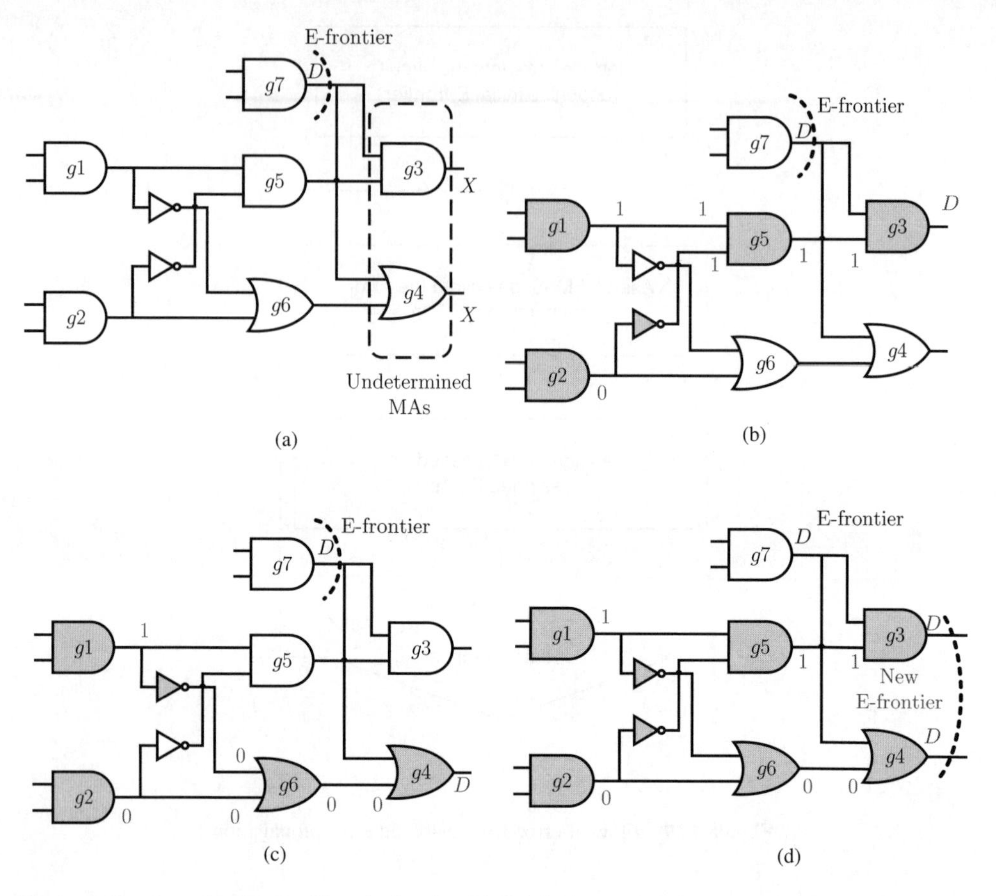

Figure 4.31 Locating new multi-node frontier. (a) Original error frontier; (b) path$_{g3}$-MA propagation; (c) path$_{g4}$-MA propagation; (d) new error frontier

On the other hand, as shown in Figures 4.31 and 4.32, when the current E-frontier node has multiple fanout nodes with undetermined mandatory assignment values (X), recursive learning has to be applied to identify the new E-frontier nodes. The definition of "path mandatory assignments" $path_{node}$-*MA*, which is used to formulate the error-frontier shifting procedure, is listed below:

Definition 4.12 *Given a Boolean network, a target wire w_t and its removal error $\epsilon_r(w_t)$, and a gate s which is in the transitive fanout cone of w_t, a gate g has path$_s$-MA value m if:*

- *V is the set of test vectors that can activate $\epsilon_r(w_t)$ and propagate it to primary outputs through paths via gate s;*
- *g has the same value m in all test vectors in V.*

Basically, Definition 4.12 states that the path$_s$-MA with value m has to be assigned to a gate g for the test of $\epsilon_r(w_t)$ if the error propagation passes through the gate s. Then, using these

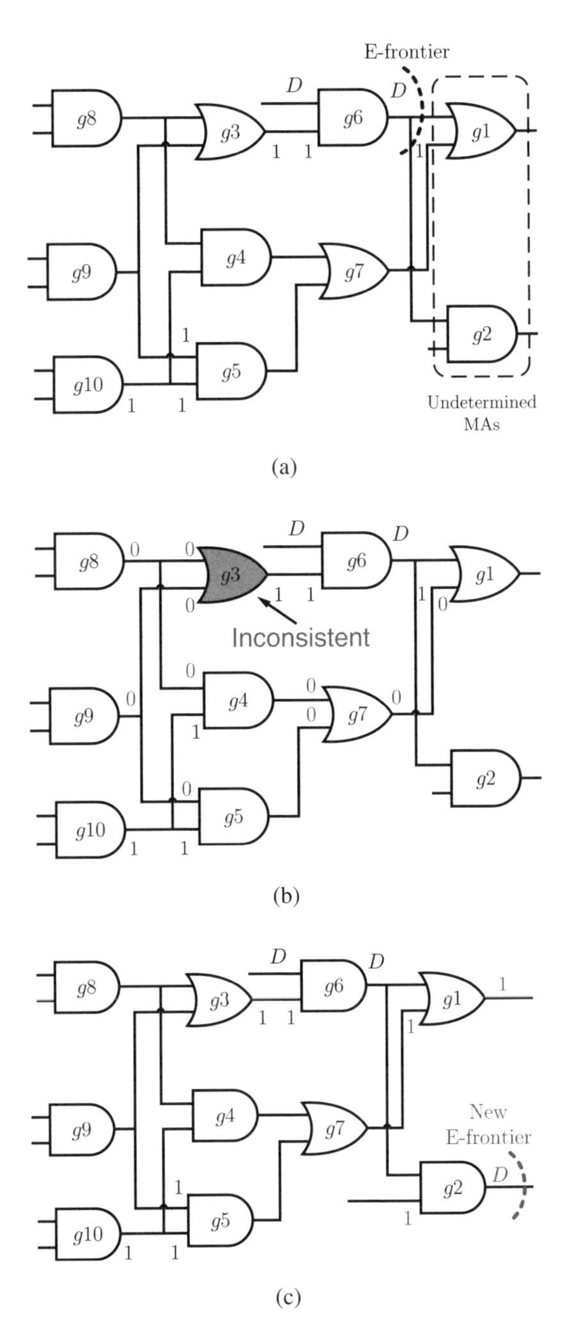

Figure 4.32 Locating new error frontier with inconsistent path$_{node}$-MA propagation. (a) Original error frontier; (b) inconsistent path$_{g1}$-MA propagation; (c) new error frontier

new MA values, the E-frontier may be further shifted toward the primary outputs. The whole propagation procedure is similar to recursive learning (Kunz and Pradhan 1994) with maximum recursive depth = 1. The definition of "path core mandatory assignments" $path_s$-CMA is constructed in accordance with $path_s$-MA.

Figure 4.31 shows an example of extending a single-node frontier to a new multi-node frontier. In Figure 4.31(a), gates $g3$ and $g4$ are the fanouts of the E-frontier and have got undetermined mandatory assignments. For each fanout gate, $path_{g3}$-MAs and $path_{g4}$-MAs are assigned, respectively, as shown in Figure 4.31(b) and (c). Similar to the mechanism of recursive learning, the intersection of the two $path_{node}$-MAs $g1 = 1, g2 = 0$ can be implied. The new multi-node frontier can then be located as shown in Figure 4.31(d).

It is not always the case that an E-frontier will keep shifting without encountering conflicts in mandatory assignments. In Figure 4.32, the propagation of the $path_{g1}$-MAs induces inconsistency at $g3$, as illustrated in Figure 4.32(b). This implies that the error cannot be propagated through $g1$ and no new multi-node frontier can be obtained. Instead, a new single-node frontier containing gate $g2$ can be located, as indicated in Figure 4.32(c).

When an error frontier contains multiple gates, all fanouts of each gate contained in the error frontier are examined and the corresponding $path_{node}$-MA propagation is performed as

Algorithm 4.9: Error frontier update procedure

```
   input : Original error frontier OE-frontier
 1 begin
 2     Gate set S = φ ;
 3     foreach gate n in OE-frontier do
 4         foreach fanout gate fn of n do
 5             if fn has no MA then
 6                 put fn into S ;
 7     foreach each gate n in S do
 8         Set uncontrolling value x as path_n-MA to the side input of n ;
 9         Perform path_n-MA propagation ;
10         if Propagation results in a conflict then
11             Set MA value x̄ to n and its side input ;
12         else
13             Record path_n-MA values of all gates ;
14     foreach each gate n in S do
15         if n is assigned different path_node-MA values with respect toeach consistent
           propagation then
16             return update error frontier failed ;
17             else
18                 S
19             et MA value to n ;
20     return update error frontier successful ;
21 end
```

mentioned to locate a new error frontier. The error frontier shifting procedure can be summarized in Algorithm 4.9.

4.4.2.4 Relationship Between Dominator-Based and Error-Frontier-Based Error Propagations

With reference to Figure 4.33(a), mandatory assignments for testing a stuck-at fault are propagated from the dominators in the traditional dominator-based error propagation. Regarding error-frontier-based error propagation, the mandatory assignments can also be propagated from the dynamic dominators, the error frontiers, and the dominators, as shown in Figure 4.33(b).

Compare the examples shown in Figures 4.26(c) and 4.28(c). The MAs of gates $g5$, $g10$, and $g11$ can be determined only by propagating the error through $g10$ and assigning the side inputs of $g10$, which is not a dominator but an error frontier. Based on this observation, the authors of CECR derived the following lemma:

Lemma 4.3 *With regard to a stuck-at fault, the set of mandatory assignments found by the error-frontier-based error propagation is a superset of that found by the dominator-based error propagation.*

4.4.3 Cut-Based Error Cancellation Rewiring

CECR is an improvement upon the FECR (Yang et al. 2012). In FECR, the rewiring process has two stages: (i) identification of the destination nodes, and (ii) identification of the source nodes of alternative wires. CECR achieves a significant speedup over FECR in the identification

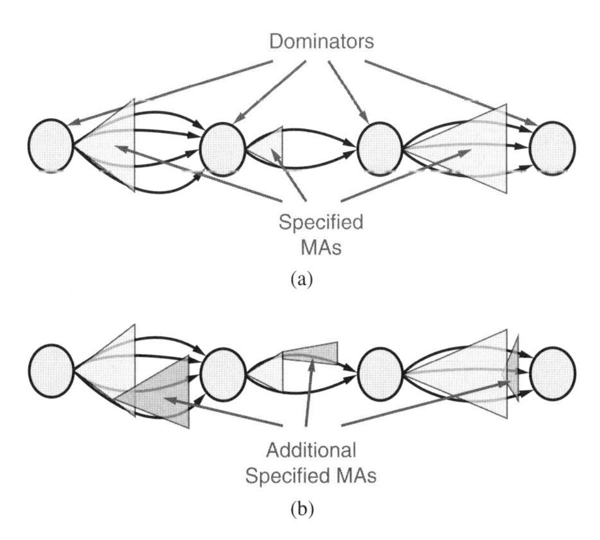

Figure 4.33 Relationship between (a) dominator-based propagation and (b) error-frontier-based propagations

process for the sinks of alternative wires. The improvement is due to the use of cut enumeration in CECR instead of network flow algorithms in FECR.

4.4.3.1 Flow of Sink Gate Identification

The addition of alternative wires can be considered as a process of injecting errors at the destination nodes of the alternative wires. Since the errors injected cancel with the error caused by the removal of the target wire, the circuit's function is preserved. A natural set of alternative wire sinks are the nodes in error cuts. In CECR, error cuts are found by an enumeration algorithm (Chan and Cong 2004, Cong et al. 1999), and each cut can be used to find a set of candidate sink nodes of the alternative wires of a given target wire. The *error graph*, which is used to identify error cuts, is defined as follows:

Definition 4.13 *Given that the node set EF is the union of all the E-frontiers (Definition 4.8), the error graph is the induced sub-graph of the original Boolean network by the node set EF, with one extra dummy node $Sink$ and virtual edges connected between each node of the last E-frontier and the $Sink$ node.*

Figure 4.34 shows an example of how an error graph is constructed. $b \to g1$ is the target wire w_t, and the error propagation of the wire removal error $\epsilon_r(w_t)$ is shown in Figure 4.34(a). After the error propagation, the following MAs are identified: $a =$

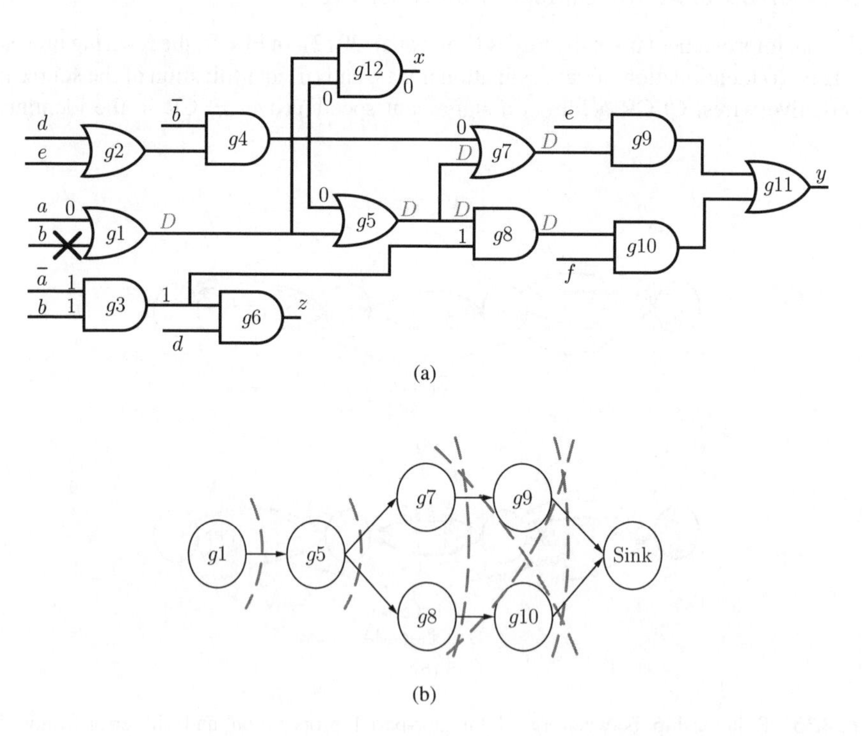

Figure 4.34 Construction of an error graph. (a) Error propagation; (b) error graph

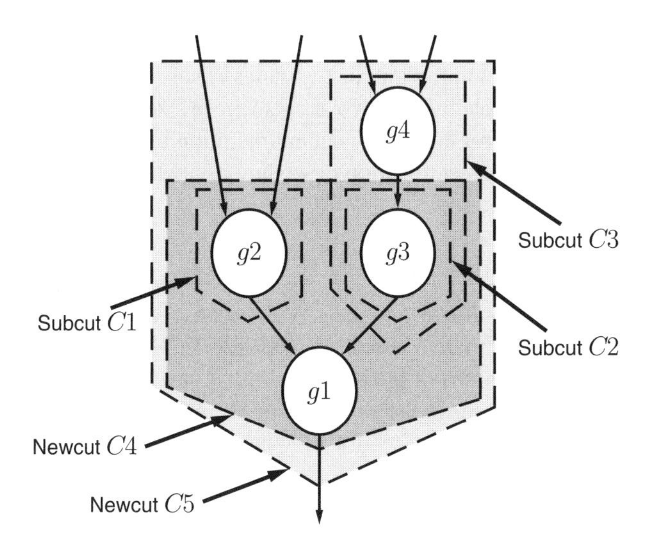

Figure 4.35 Cut enumeration

$0, b = D, g1 = D, g3 = 1, g4 = 0, g12 = 0, g5 = D, g7 = D, g8 = D$. Then the path$_{g9}$-MAs are $e = 1, g9 = D$, and the path$_{g10}$-MAs are $f = 1, g10 = D$. Finally, an error graph including all gates with error-MAs is obtained as shown in Figure 4.34(b), where each dotted line represents a different error cut.

Obviously, any cut in an error graph is an error cut for the propagation of the error $\epsilon_r(w_t)$. Therefore, a cut enumeration algorithm can be used to find all K-feasible error cuts for a given K. The cut enumeration to find all K-feasible cuts rooted at a node v can be guided by the following Equation (Cong et al. 1999, Chan and Cong 2004):

$$f(K, v) = \otimes^K_{u \in input(v)}[u + f(K, u)] \tag{4.7}$$

where $f(K, v)$ represents all the K-feasible cuts rooted at node v, the operator $+$ represents Boolean OR, and $\otimes^K$ is Boolean AND on its operands, excluding all cuts of over K gates.

The cut enumeration process is illustrated in Figure 4.35. To enumerate all cuts rooted at $g1$, all cuts rooted at its fanins, $g2$ and $g3$, are first built as subcuts $C1$ at $g2$ and $C2$, $C3$ at $g3$. Each of the subcuts $C1$, $C2$, and $C3$ is then merged with $g1$ to form a new cut. By merging $C1$ with $g1$ and then $C2$, the new cut $C4$ can be constructed. Similarly, another cut of $g1$, $C5$, can be constructed by merging $C1$, $C3$, and $g1$.

In CECR, the cut enumeration algorithm is employed to find all cuts of each gate of the error graph in a topological order from the target wire to the virtual $Sink$.

4.4.3.2 Complexity of Sink Identification

The complexity of the cut enumeration method is proportional to the number of cuts on a gate, which is $O(n^K)$ for the worst case, where K is the maximum cut size. However, the authors in (Chan and Cong 2004) found that when the input K is small ($K < 6$), the number of cuts generated for each gate is a small constant. Since the authors in FECR found that using error cuts of size not larger than 2 is sufficient to remove almost any wire, the authors of CECR also considered only error cuts containing one or two gates. Their experiments show that almost any wire in a circuit is removable by using error cuts whose sizes are smaller than 2. Thus the complexity of collecting all error cuts using cut enumeration in CECR is just $O(n)$, which is also the complexity of the sink identification procedure as a whole.

4.4.3.3 Identification of Source Gates

After identifying the set of candidate sink gates of alternative wires, the next task is to figure out the source gates of the alternative wires. As in IRRA, ECR, and FECR, rectification networks are first built as the sources of alternative wires. Then, the rectification networks are substituted by some existing gates in the circuit using the node merging algorithm (Chen and Wang 2010a). In CECR, the concept of rectification networks is extended to correct also errors at path_{node}-error gates in error cuts.

The logical conjunction of CMAs and path_{node}-CMAs is denoted by $AND(CMA, path_t\text{-}CMA)$. It defines the test vectors that can propagate the error via node t. When a path_{node}-error gate is taken as a potential destination node of an alternative wire, $AND(CMA, path_t\text{-}CMA)$ is used as the rectification network. We will use the term *rectification network(s)* to represent both $AND(CMA)$ and $AND(CMA, path_{node}\text{-}CMA)$ in the following text.

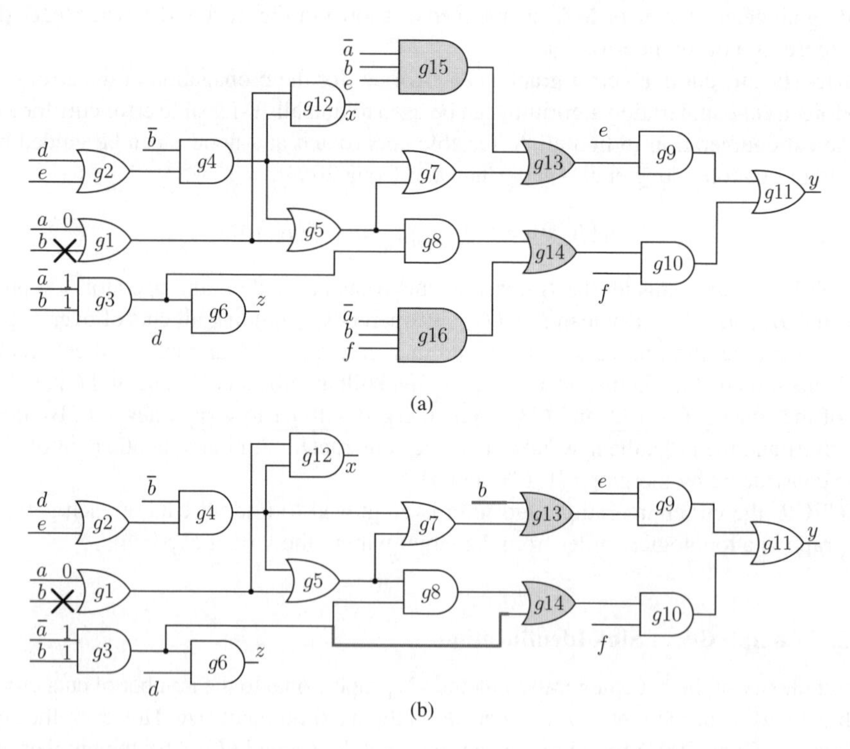

Figure 4.36 Simplification of rectification networks by node merging. (a) Rectification networks at $g7$ and $g8$; (b) rewired circuit

As example showing how rectification networks are constructed is shown in Figure 4.34(a). In the figure, CMAs are collected as $a = 0, b = 1$. Path$_{node}$-CMA is $e = 1$ with respect to gate $g7$, while it is $f = 1$ with respect to gate $g8$. Then, the rectification network AND $(CMA, path_{node}\text{-}CMA)$ is $\bar{a} \cdot b \cdot e$ with respect to $g7$, and it is $\bar{a} \cdot b \cdot f$ with respect to $g8$.

Once the rectification networks are built to cancel the error caused by target wire removal, a node merging algorithm proposed in NAR (Chen and Wang 2009, 2010a, b, 2012) is performed to try to replace each of the rectification networks by a single existing gate or an additional logic consisting of two existing gates in the original circuit.

The procedure of rectification network simplification is shown in Figure 4.36. Two rectification networks are built at $g7$ and $g8$, respectively, as shown in Figure 4.36(a). The node merging algorithm is used twice to merge $g15$ with b, and $g16$ with $g3$. The successfully rewired circuit is shown in Figure 4.36(b).

The overall framework of CECR is shown in Algorithm 4.10.

Algorithm 4.10: CECR framework

input : target wire w_t and its removal error w_t_error

1 **begin**
2 Identify all MAs (illustrated in Section 4.4.2) ;
3 Set error cut set $S = \phi$;
4 Collect all error cuts into S using cut enumeration (Cong et al. (1999)) ;
5 **foreach** *error cut E-cut in S* **do**
6 **foreach** *gate n in E-cut* **do**
7 Build rectification network for n ;
8 **if** *rectification network can be merged into existing gates (Chen and Wang (2012))* **then**
9 Find AW(s) to block w_t_error going through n;
10 **if** *find AW(s) to block w_t_error going through each gate of E-cut* **then**
11 Find available rewiring solution consisting of AW(s) ;
12 **end**

4.4.4 Verification of Alternative Wires

If there is only one gate in the error cut, the alternative wire is always valid. Otherwise, validation of the alternative wire is required to ensure that the function of the rewired circuit is correct. The following lemma indicates the condition when validation is necessary:

Lemma 4.4 *Suppose there are two gates u and v in the same error cut. The corresponding path-MAs for each of gate are found by propagating the target wire removal error $\epsilon_r(w_t)$ through u and v, respectively. If $\epsilon_r(w_t)$ is to be corrected at both gates u and v at the same time, the necessity of verification is based on the following conditions:*

- *If $AND(path_u\text{-}CMA) \cap AND(path_v\text{-}CMA) = \phi$, verification can be exempted.*
- *If $AND(path_u\text{-}CMA) \cap AND(path_v\text{-}CMA) \neq \phi$, verification is necessary.*

Proof. Suppose that the intersection is empty. Since $AND(path_u\text{-}CMA)$ represents the necessary condition for all the test vectors that can distinguish between the good circuit and rewired circuit when the error $\epsilon_r(w_t)$ is propagated through u, it is impossible for $AND(path_v\text{-}CMA)$ to affect the error propagation through u. The two test vectors cannot be effective simultaneously, and therefore a verification after rewiring is not necessary. ■

For instance, verification is necessary if error $\epsilon_r(b \rightarrow g1)$ is rectified at $g7$, $g8$ in the example shown in Figure 4.34.

4.4.5 Complexity Analysis of CECR

Assume that the number of gates in a circuit is n. The complexity of sink gate identification is $O(n)$, as reported in previous section, and the complexity of source gate identification is $O(n)$, as reported in Yang et al. (2012). Hence, the complexity of identifying all sources at all potential sinks of alternatives is $|g_{sinks}| \cdot O(n) = O(n^2)$. The total time complexity of CECR is then $O(n^2)$, which is the same as in RAR and ECR. It is much smaller than that of FECR ($O(n^5)$). It has been reported that worse scenarios rarely happen, and the practical complexity of calling CECR once is found to be much lower, as shown in Table 4.11.

4.4.6 Relationship Between ECR, FECR, and CECR

An error cut is a set of gates such that the target wire removal error can be cancelled before it propagates through these gates. If there is only one gate in an error cut, the gate is essentially a dominator or dynamic dominator. Therefore, the circuit transformations that can be identified by ECR are a subset of FECR or CECR. In CECR, a new error propagation scheme is proposed to find more MAs that are a superset of that used in FECR, as shown in Section 4.4.2. Thus the error cuts found by CECR form a superset of FECR. Therefore, CECR is a superset of FECR.

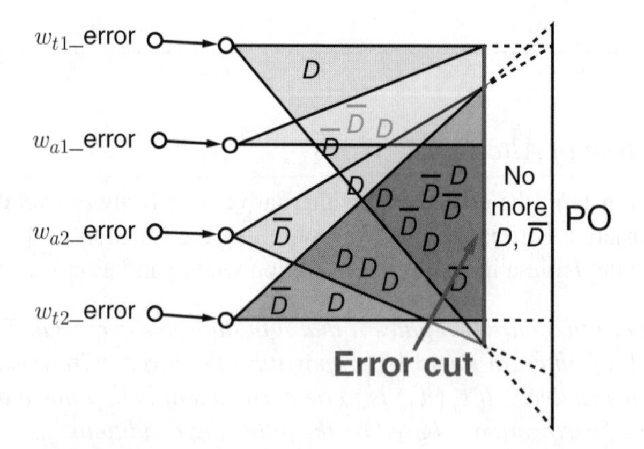

Figure 4.37 Structural view of n-to-m error cancellation

4.4.7 *Extending CECR for n-to-m Rewiring*

Suppose that there are n target wires (represented by $w_{t1}(i_1, j_1)$ to $w_{tn}(i_n, j_n)$). Each target wire removal induces an error (represented by w_{t1-} error to w_{tn-} error) into the circuit. If we can prevent all these errors from being observed at any primary outputs, then all target wires can be removed altogether while preserving the original circuit's function. The structural view of n-to-m error cancellation is illustrated in Figure 4.37.

In order to remove all target wires, in the CECR flow all destination nodes of the target wires are collected (from g_{j_1} to g_{j_n}) to form an initial error frontier. Then, an error-frontier-based error propagation is performed. All error cuts are collected to identify the destination nodes of alternative wires. Finally, the sources of the alternative wires are identified using the flow discussed in Section 4.4.3.3. Based on this discussion, it can be seen that CECR can be naturally extended to handle multiple target wires simultaneously.

Algorithm 4.11: Sub-circuit creation

```
   input : target wire w_t, window size L, and fanout limit K
1  begin
2      Set queue q = φ;
3      Set sub-circuit set S = φ;
4      Push source and sink gates of w_t into q and S;
5      while (q ≠ φ)&(|S| < L) do
6          Pop up gate n from the head of q;
7          if |fanout_n| ≥ K then
8              continue;
9          foreach fanin gate fn of n do
10             if fn is not in S then
11                 Push fn into q and S;
12         foreach fanout gate fn of n do
13             if fn is not in S then
14                 Push fn into q and S;
15     foreach gate n in S do
16         foreach each fanin gate fn of n do
17             if fn is not in S then
18                 Create virtual primary input vi to replace fn driving n;
19         foreach fanout gate fn of n do
20             if fn is not in S then
21                 Create virtual primary output vo to replace fn driven by n;
22 end
```

4.4.8 *Speedup for CECR*

Most logic restructuring techniques suffer from high CPU usage. As an example, the authors of Plaza and Markov (2008) proposed an algorithm that combines logic restructuring techniques and physical synthesis techniques to reduce circuit delay. The algorithm requires more than 1000 s to optimize a circuit with about 7000 cells. This shows that the algorithm is not quite scalable and may not be able to handle large-scale industrial designs having millions of standard cells.

The authors of CECR introduced a windowing technique for logic rewiring to save CPU time efficiently, especially for large circuits. Their idea was to process only a small region of the circuit at a time. In the exact implementation, a sub-circuit (window) of certain *window size L* is created around a target wire by means of breadth-first search (BFS). That is to say, the sub-circuit for a target wire contains the source and destination nodes of the target wire initially. It is then expanded by adding its neighbors recursively until it attains the given window size. The window size is defined as the maximum number of gates contained in the sub-circuit.

Figure 4.38(a) depicts the concept of a sub-circuit. After defining the boundary of the sub-circuit, virtual primary inputs and outputs are created temporarily to replace the gates lying outside the boundary, as shown in Figure 4.38(b). The procedure for the creation of sub-circuits is listed in Algorithm 4.11.

It was experimentally observed by CECR's authors that propagating errors through gates with extremely large fanouts would require much CPU time, but could obtain only a few alternative wires. Therefore, they suggested the exclusion of all gates with more than a certain number of fanouts in the sub-circuit construction process.

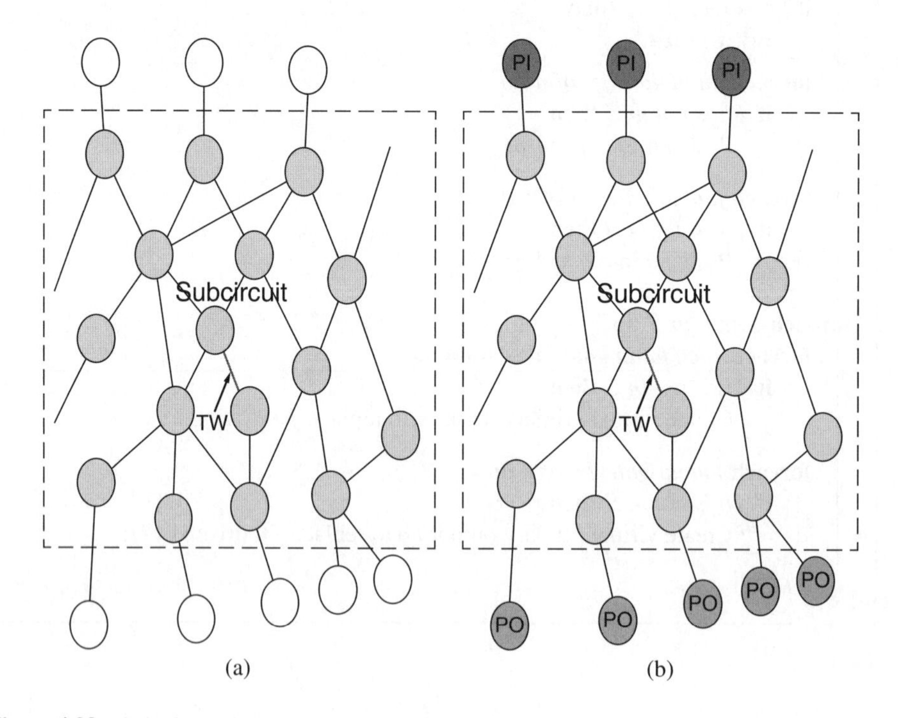

Figure 4.38 Sub-circuit for rewiring. (a) Collect gates for sub-circuit; (b) create virtual primary IO

4.4.9 *Experimental Results*

The CECR scheme was implemented in C++ and was evaluated under the same computing environment as used in FECR (with one 2.8-GHz CPU and 1 GB RAM) for a fair comparison on runtime efficiency. The benchmarks used were also ISCAS and MCNC circuits and were optimized by ABC first. The functions of the rewired circuit were verified by a powerful SAT solver miniSAT (Een and Sorensson 2003).

4.4.9.1 ECR versus FECR versus CECR

Table 4.8 shows the comparison between ECR, FECR, and CECR. For each circuit, all wires driving functional gates were considered target wires. The number of target wires processed during the experiments is listed in column 2 of the table. Each of the rewiring schemes is applied on every target wire one by one to find the alternative wires. The alternative wires identified were not added to the circuits, and the circuits were not restructured in any way. The columns "$r.TW\%$" show the rewiring rate (the percentage of target wires that are removable) of each rewiring scheme, respectively. It was found that ECR could replace only 33.7% wires on average, while FECR and CECR could replace almost any wire (94.8% and 96.1% respectively). The reason is that ECR identifies only a single alternative wire to replace a target wire.

Table 4.8 Comparison between ECR, FECR, and CECR

			ECR		FECR		CECR	
Benchmark	#gate	$\#TW$	*$r.TW\%$[a]	**Time (ms)[b]	$r.TW\%$	Time (ms)	$r.TW\%$	Time (ms)
9sym-hdl	49	96	27.1	0.45	81.3	72.9	85.4	2.1
pcler8	71	126	23.0	1.04	80.9	69.0	81.7	10.3
f51m	98	190	55.3	2.88	98.9	963.2	98.9	36.8
comp	93	184	37.5	1.67	74.5	1019.6	91.8	15.2
5xp1	93	178	52.2	2.71	98.9	791.6	98.9	37.1
b9_n2	93	166	42.8	0.66	94.6	164.5	94.6	4.2
my_adder	129	256	43.8	0.26	86.7	105.1	86.7	1.2
ttt2	143	274	47.8	1.84	95.3	272.6	97.1	16.1
term1	159	314	51.3	2.79	95.5	1239.2	97.1	25.8
sao2-hdl	162	324	46.3	5.88	92.9	2413.6	92.9	111.7
C432	166	320	46.3	2.73	91.2	297.2	91.3	13.8
C1908	355	706	25.5	3.25	97.0	694.5	98.4	30.3
C880	340	648	32.4	1.25	94.1	400.2	95.2	8.5
C1355	418	772	22.3	1.64	98.4	548.1	100.0	24.0
rot	502	942	36.8	1.30	93.6	791.8	94.1	7.2
x3	617	1120	26.2	3.54	97.2	968.8	98.3	39.5
apex6	673	1188	30.2	3.81	97.5	1670.0	98.6	40.5
Average			33.7	2.55	94.8	885.4	96.1	27.7
Time ratio				0.09×		32×		1×

[a]$r.TW\%$: Percentage of *removable* target wires.
[b]Time (ms): CPU time of trying to remove one target wire.

Table 4.9 Detailed statistics for CECR

Benchmark	*TW*		1-to-1[a]		1-to-2[b]		1-to-3+		FAIL[c]	
	#	%	#	%	#	%	#	%	#	%
9sym-hdl	96	100	26	27.1	54	56.3	2	2.1	14	14.6
pcler8	126	100	29	23.0	74	58.7	0	0.0	23	18.3
f51m	190	100	105	55.3	83	43.7	0	0.0	2	1.1
comp	184	100	69	37.5	96	52.2	4	2.2	15	8.2
5xp1	178	100	93	52.2	83	46.6	0	0.0	2	1.1
b9_n2	166	100	71	42.8	83	50.0	3	1.8	9	5.4
my_adder	256	100	112	43.8	110	43.0	0	0.0	34	13.3
ttt2	274	100	131	47.8	134	48.9	1	0.4	8	2.9
term1	314	100	161	51.3	134	42.7	10	3.2	9	2.9
sao2-hdl	324	100	147	45.4	152	46.9	2	0.6	23	7.1
C432	320	100	148	46.3	135	42.2	9	2.8	28	8.8
C1908	706	100	180	25.5	496	70.3	19	2.7	11	1.6
C880	648	100	210	32.4	404	62.3	3	0.5	31	4.8
C1355	772	100	172	22.3	584	75.6	16	2.1	0	0.0
rot	942	100	346	36.7	524	55.6	16	1.7	56	5.9
x3	1120	100	293	26.2	792	70.7	16	1.4	19	1.7
apex6	1188	100	359	30.2	798	67.2	14	1.2	17	1.4
Total	7804	100	2652	34.0	4736	60.6	115	1.5	301	3.9%

[a] 1-to-1: Number (percentage) of target wires that can be replaced by a single alternative wire.
[b] 1-to-2(3+): Number (percentage) of target wires that can be replaced by at least 2(3+) alternative wires.
[c] FAIL: Number (percentage) of target wires that cannot be replaced by any alternative wire.

However, FECR and CECR can construct more complex alternative logic to replace target wires. With regard to CPU time, CECR was much faster to process one target wire, compared to FECR. It was 31.9× faster than FECR.

4.4.9.2 Detailed Statistics for CECR

Table 4.9 shows the detailed statistics for CECR. Columns "1-to-1/2/3+" show the number and percentage of target wires that can be replaced by at least 1/2/3+ alternative wires. Around 94.6% of target wires could be replaced by no more than two alternative wires (34.0 + 60.6%). In certain applications, using two alternative wires to substitute one target wire does not involve a very high cost and may be beneficial. Roughly 1.5% of target wires required three or more alternative wires to replace. There was 3.9% target wires whose alternative wires could not be determined and therefore were not removable.

4.4.9.3 Effectiveness of Windowing Technique

The effectiveness of the windowing technique adopted in CECR was also tested. Larger benchmarks from the IWLS 2005 benchmark suite were used to evaluate the technique. In the experiments, all logic blocks with multiple fanins in the circuits were decomposed into two-input

Table 4.10 Effectiveness of windowing technique

Benchmark	#gate	$\#TW$	$r.TW\%$	Time (ms)
s38417	9,211	18,420	76.7	2.3
systemcaes	12,445	24,888	73.2	1.7
ac97_ctrl	14,271	28,538	65.2	1.1
usb_funct	15,936	31,908	73.1	2.8
mem_ctrl	16,057	32,110	63.2	4.6
b21	20,710	41,418	76.2	3.8
aes_core	21,717	43,432	76.2	2.7
DMA	24,614	49,224	70.3	2.3
b22	31,400	62,798	76.2	4.0
Ethernet	86,745	173,486	63.9	2.5
Average			71.4	2.8

blocks. All buffers were removed, and all inverter blocks were merged with other logic blocks. The benchmarks were made into combinational sub-circuits by considering flip-flops as PIs and POs.

The number of tested wires is shown in column "$\#TW$" of Table 4.10. For each benchmark, all wires driving logic blocks are regarded as target wires. The alternative wires for each target wire were determined inside a sub-circuit whose window size was 100 (Section 4.4.8). All gates that have more than 10 fanouts were not processed during sub-circuit construction. The rewiring rate and average CPU time consumed to process one target wire are shown in columns "$r.TW\%$" and "time (ms)." It is obvious that using the windowing approach could reduce rewiring rate (the average rate was only 71.4% compared to 96.1% as reported in Table 4.8).

Table 4.11 Effectiveness of different windowing sizes

Parameters[a]		$r.TW\%$	Time
L	K		(ms)
50	5	68.0	1.7
100	10	76.2	4.0
150	15	80.0	6.7
200	20	81.8	9.0
300	30	82.9	13.4
400	40	83.4	19.0
500	50	83.8	24.1
600	60	84.3	28.6
700	70	85.7	43.0
800	80	86.4	55.2
∞[b]	∞	96.8	18,024.5

[a] L is the window size and K is the fanout limit.
[b] No windowing technique applied.

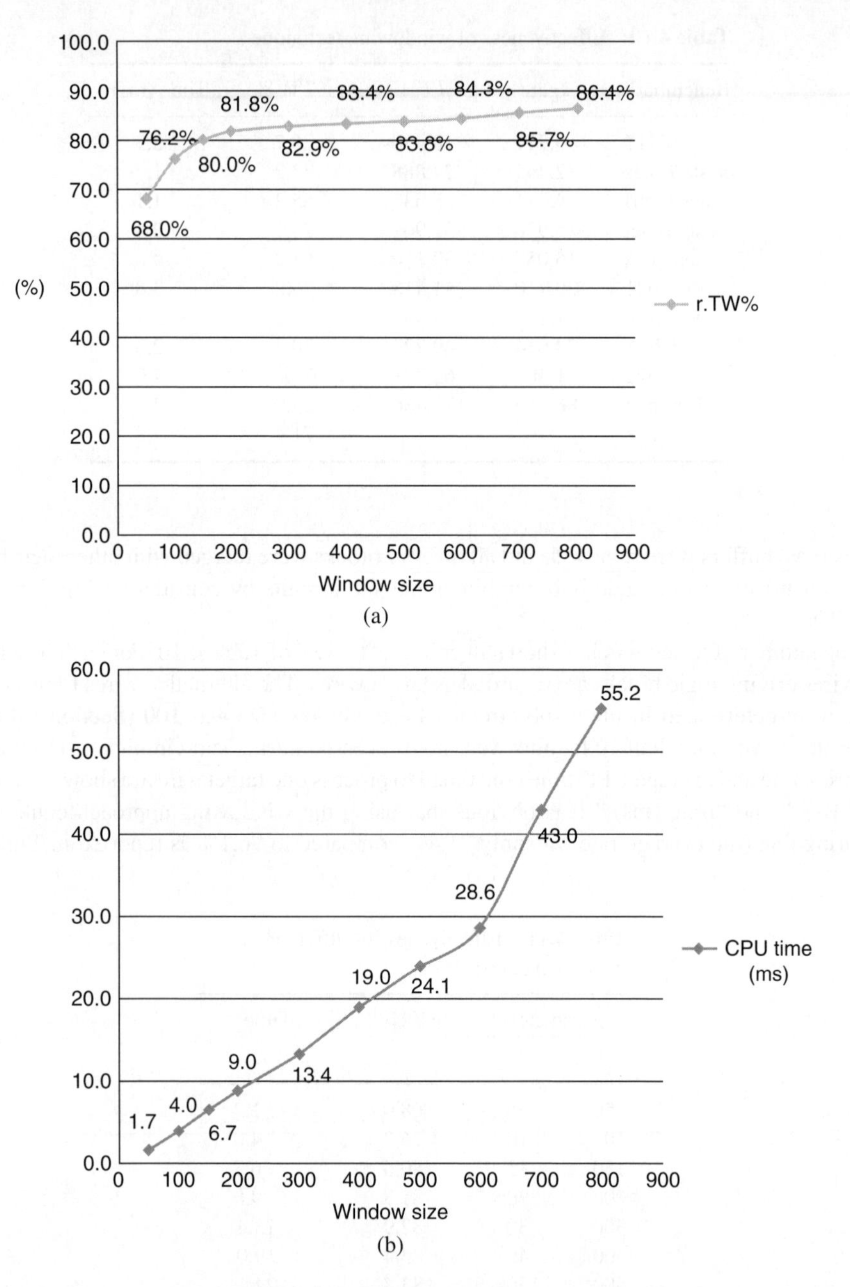

Figure 4.39 Rewiring ability of CECR for different window sizes. (a) Rewiring rates for different window sizes; (b) CPU times for different window sizes

However, it is a tradeoff, and using the windowing technique could constrain the CPU time within a small constant value (about 2.8 ms on average) almost regardless of circuit size. For large-scale industrial benchmarks, it is worth trading rewiring rate for speed.

The authors of CECR used the benchmark `b22` to test CECR under different window sizes and fanout limits. Table 4.11 lists the statistics in great detail. It can be observed that the rewiring rate has positive correlation with the window size. The CPU time needed to find the alternative wires for one wire also increases directly with the window size. The last row of the table shows the performance of CECR when no windowing technique was applied. Although a very high rewiring rate (more than 96%) can be obtained, it takes around 10 s to process just one wire.

Figure 4.39(a) shows the relationship between rewiring rate and window size. We can see that the rewiring rate shows very little improvement when the window size is larger than 200. The CPU time required under different window sizes is shown in Figure 4.39(b). It is clear that the CPU time increases more and more rapidly when the window size is larger. From this statistics, we can probably claim that using a window size of 200 in CECR for large circuits is a good choice.

It is not a secret that in today's design process, a quite long time is often spent on the very last stage where there are a few dozens or hundreds of routing requirements that cannot be satisfied even with repeated iterations of physical synthesis optimizations. In this chapter, the latest error- cancellation-based rewiring scheme CECR was discussed. By adding single or multiple wires to a circuit, almost any (96%) unwanted single wire existing in the circuit can be removed. The worst case runtime complexity of CECR is $O(n^2)$, which is far better than that of FECR whose complexity is $O(n^5)$. Empirical studies have shown that CECR could achieve 32 times speedup over FECR. This significant progress makes rewiring a more practical logic synthesis technique.

References

Berkeley Logic Synthesis and Verification Group. ABC: A system for sequential synthesis and verification, release 70911. URL http://www.eecs.berkeley.edu/ alanmi/abc.

D. Chan and J. Cong. DAOmap: a depth-optimal area optimization mapping algorithm for FPGA designs. In *Proceedings of the ICCAD*, pages 752–759, November 2004.

S. C. Chang and M. Marek-Sadowska. Perturb and simplify: multilevel Boolean network optimizer. *IEEE Transactions on Computer-Aided Design of Integrated Circuits and Systems*, 15(12):1494–1504, 1996.

C.-W. J. Chang and M. Marek-Sadowska. Who are the alternative wires in your neighborhood? (alternative wires identification without search). In *Proceedings of the 11th Great Lakes symposium on VLSI*, GLSVLSI '01, pages 103–108, New York, 2001. ACM. ISBN: 1-58113-351-0. doi: 10.1145/368122.368880.

C. W. J. Chang and M. Marek-Sadowska. Theory of wire addition and removal in combinational Boolean networks. *Microelectronic Engineering*, 84(2):229–243, 2007.

S.-C. Chang, L. P. P. P. Van Ginneken, and M. Marek-Sadowska. Circuit optimization by rewiring. *IEEE Transactions on Computers*, 48(9):962–970, 1999. ISSN: 0018-9340. doi: 10.1109/12.795224.

Y.-C. Chen and C.-Y. Wang. Fast detection of node mergers using logic implications. In *Computer-Aided Design - Digest of Technical Papers, 2009. ICCAD 2009. IEEE/ACM International Conference on*, pages 785–788, November 2009.

Y.-C. Chen and C.-Y. Wang. Node addition and removal in the presence of don't cares. In *Design Automation Conference (DAC), 2010 47th ACM/IEEE*, pages 505–510, 2010a.

Y.-C. Chen and C.-Y. Wang. Fast node merging with don't cares using logic implications. *IEEE Transactions on Computer-Aided Design of Integrated Circuits and Systems*, 29(11), 2010.

Y.-C. Chen and C.-Y. Wang. Logic restructuring using node addition and removal. *IEEE Transactions on Computer-Aided Design of Integrated Circuits and Systems*, 31(2), 2012.

K.-T. Cheng and L. A. Entrena. Multi-level logic optimization by redundancy addition and removal. In *Design Automation, 1993, with the European Event in ASIC Design. Proceedings. [4th] European Conference on*, pages 373–377, February 1993. doi: 10.1109/EDAC.1993.386447.

J. Cong, C. Wu, and Y. Ding. Cut ranking and pruning: enabling a general and efficient FPGA mapping solution. In *Proceedings of Field-Programmable Gate Arrays*, 1999.

M. Damiani and G. De Micheli. Observability don't care sets and boolean relations. *Computer-Aided Design, 1990. ICCAD-90. Digest of Technical Papers., 1990 IEEE International Conference on*, pages 502–505, November 1990. doi: 10.1109/ICCAD.1990.129965.

N. Een and N. Sorensson. An extensible SAT-solver. In *Proceedings of SAT*, pages 502–518, 2003.

L. A. Entrena and K.-T. Cheng. Combinational and sequential logic optimization by redundancy addition and removal. *IEEE Transactions on Computer-Aided Design of Integrated Circuits and Systems*, 14(7):909 916, 1995. ISSN: 0278-0070. doi: 10.1109/43.391740.

R. Krieger, B. Becker, R. Hahn, and U. Sparmann. Structure based methods for parallel pattern fault simulation in combinational circuits. In *Proceedings of the European Conference on Design Automation. EDAC.*, pages 497–502. IEEE, 1991.

W. Kunz and D. K. Pradhan. Recursive learning: a new implication technique for efficient solutions to CAD problems-test, verification, and optimization. *IEEE Transactions on Computer-Aided Design of Integrated Circuits and Systems*, 13:1143–1158, 1994.

T.-K. Lam, W.-C. Tang, X. Yang, and Y.-L. Wu. ECR: a powerful and low-complexity error cancellation rewiring scheme. *ACM Transactions on Design Automation of Electronic Systems*, 17(4):50:1–50:21, 2012. ISSN: 1084–4309. doi: 10.1145/2348839.2348854.

C.-C. Lin and C.-Y. Wang. Rewiring using irredundancy removal and addition. In *Design, Automation Test in Europe Conference Exhibition, 2009. DATE '09*, pages 324–327, April 2009.

A. Mishchenko, S. Chatterjee, and R. Brayton. DAG-aware AIG rewriting a fresh look at combinational logic synthesis. In *DAC '06: Proceedings of the 43rd Annual Design Automation Conference*, pages 532–535, New York, 2006. ACM. ISBN: 1-59593-381-6. doi: 10.1145/1146909.1147048.

S. Muroga, Y. Kambayashi, H. C. Lai, and J. N. Culliney. The transduction method-design of logic networks based on permissible functions. *IEEE Transactions on Computers*, 38:1404–1424, 1989. ISSN: 0018-9340. doi: 10.1109/12.35836.

S. M. Plaza, K.-H. Chang, I. L. Markov, and V. Bertacco. Node mergers in the presence of don't cares. In *ASP-DAC '07: Proceedings of the 2007 Asia and South Pacific Design Automation Conference*, pages 414–419, Washington, DC, 2007. IEEE Computer Society. ISBN: 1-4244-0629-3. doi: 10.1109/ASPDAC.2007.358021.

S. M. Plaza and I. L. Markov. Optimizing nonmonotonic interconnect using functional simulation and logic restructuring. *IEEE Transactions on Computer-Aided Design of Integrated Circuits and Systems*, 27(12):2107–2119, 2008.

E. M. Sentovich, K. J. Singh, L. Lavagno, C. Moon, R. Murgai, A. Saldanha, H. Savoj, P. R. Stephan, R. K. Brayton, and A. Sangiovanni-vincentelli. SIS: a system for sequential circuit synthesis. Technical report UCBERL M92/41, 1992.

P. Tafertshofer, A. Ganz, and K. J. Antreich. Igraine-an implication graph-based engine for fast implication, justification, and propagation. *IEEE Transactions on Computer-Aided Design of Integrated Circuits and Systems*, 19(8):907–927, 2000. ISSN: 0278-0070. doi: 10.1109/43.856977.

W.-C. Tang, W.-H. Lo, and Y.-L. Wu. Further improve excellent graph-based FPGA technology mapping by rewiring. In *Proceedings of IEEE International Symposium on Circuits and Systems*, pages 1049–1052. 2007.

X. Wei, T.-K. Lam, X. Yang, W.-C. Tang, Y. Diao, and Y.-L. Wu. Delete and Correct (DaC): an atomic logic operation for removing any unwanted wire. In *VLSI Design*. IEEE Computer Society, 2013.

Z. Z. Wu and S. C. Chang. Multiple wire reconnections based on implication flow graph. *ACM Transactions on Design Automation of Electronic Systems (TODAES)*, 11(4):939–952, 2006. ISSN: 1084-4309.

X. Yang, T.-K. Lam, and Y.-L. Wu. ECR: a low complexity generalized error cancellation rewiring scheme. In *DAC '10: Proceedings of the 47th Design Automation Conference*, pages 511–516, New York, 2010. ACM. ISBN: 978-1-4503-0002-5. doi: 10.1145/1837274.1837400.

X. Yang, T.-K. Lam, W.-C. Tang, and Y.-L. Wu. Almost every wire is removable: A modeling and solution for removing any circuit wire. In *Design, Automation Test in Europe Conference Exhibition (DATE), 2012*, pages 1573–1578, 2012. doi: 10.1109/DATE.2012.6176723.

Q. Zhu, N. Kitchen, A. Kuehlmann, and A. Sangiovanni-Vincentelli. SAT sweeping with local observability don't-cares. In *DAC '06: Proceedings of the 43rd annual Design Automation Conference*, pages 229–234, New York, 2006. ACM. ISBN: 1-59593-381-6. doi: 10.1145/1146909.1146970.

5

Applications

5.1 Area Reduction

Area reduction is very important in integrated circuit (IC) design. Many operations throughout the whole design flow, including technology mapping, verification, and placement, can consequently become less complicated when the netlists are smaller.

Node merging is an important technique for area reduction. It reduces the circuit area by merging the logically equivalent nodes and thus reduces the number of nodes in a netlist. Rewriting Mishchenko et al. (2006) and sweeping Kuehlmann (2004) are the earliest node-merging algorithms known. They are effective and can be easily scaled for large circuits. But the constraint of these algorithms is that two nodes must be logically equivalent to be merged. This requirement is too strict and prevents us from finding more merging solutions.

Recently, some improved node-merging algorithms (Zhu et al. 2006, Plaza et al. 2007,Chen and Wang 2009, 2010) have been proposed. Their advantage over the earlier node-merging methods is that these algorithms explore the concepts of observability don't cares (ODCs). By utilizing the information of ODCs, two nodes that are not logically equivalent can still be merged if their difference is included in one of the node's ODCs.

In Plaza et al. (2007), the authors proposed a Boolean satisfiability-based (SAT-based) node-merging algorithm that computes the approximate ODCs of each node and applies SAT to check the correctness. NAR (Chen and Wang 2009) is an automatic test pattern generation (ATPG)-based algorithm, which replaces a target node by another existing node inside the netlist. In Chen and Wang (2010), NAR was improved to allow replacement of a target node by an additional new node when no suitable existing nodes could be found for substitution. The two different methods in Plaza et al. (2007) and Chen and Wang (2010) have shown comparable area reduction capability when tested under the same benchmarks. With regard to speed, Chen and Wang (2010) runs faster because ATPG-based techniques are intrinsically more efficient than SAT-based ones.

We observe that both algorithms highly depend on the initial ODC distributions of the circuits and the results can be further improved if ODC information can be redistributed following certain guidelines. Suppose there is a target node having no substitute nodes. If we can change the logic function of the target node, or enlarge the Boolean space of its ODCs

Boolean Circuit Rewiring: Bridging Logical and Physical Designs, First Edition.
Tak-Kei Lam, Wai-Chung Tang, Xing Wei, Yi Diao and David Yu-Liang Wu.

without affecting any function of the primary outputs (POs), we may have a higher chance of replacing that target node.

To explore the above idea, in our work, we propose a new area reduction algorithm in which node-based and wire-based logic synthesis algorithms are coupled. When the node-based algorithm fails to find any merging solution for a target node, we use the wire-based algorithm to modify its logic function or ODC toward the direction of enhancing its chance of being merged while maintaining the functionality of the entire circuit. In this work, we choose NAR Chen and Wang (2010) as the basic node-based method and ECR (Yang et al. 2010), which is a rewiring technique, as the wire-based algorithm.

The effectiveness of this scheme is justified by empirical results, which demonstrate nearly double area reduction over all other previous works.

5.1.1 Preliminaries

In this section, we will introduce some background knowledge about our work. We first explain the definition of ODC, and then discuss the node-merging algorithm NAR. Finally, we will explain the principle of the rewiring engine ECR.

5.1.1.1 Observability Don't Cares (ODCs)

ODCs occur when the value of an internal node does not affect the outputs of the circuit because of limited observability. For example, in Figure 5.1, $a = 0$ implies $n_2 = 0$. Since $n_2 = 0$ is an input-controlling value of n_4, it prevents the value of n_3 from being observed at n_4. Then, the output value of node n_3 is a don't care under such a situation.

Suppose the logic function of the target node n_t and substitute node n_s are $F(n_t)$ and $F(n_s)$, respectively. The ODC of n_t is represented as $ODC(n_t)$. With the flexibility of ODCs, $F(n_t)$ and $F(n_s)$ are not required to be equivalent so long as Equation 5.1 is satisfied.

$$F(n_t) \oplus F(n_s) \subset ODC(n_t) \tag{5.1}$$

For example, in Figure 5.1, n_1 and n_3 are not functionally equivalent. However, the values of n_1 and n_3 only differ when $a = b$. Additionally, $a = b$ implies $n_2 = 0$. As discussed previously,

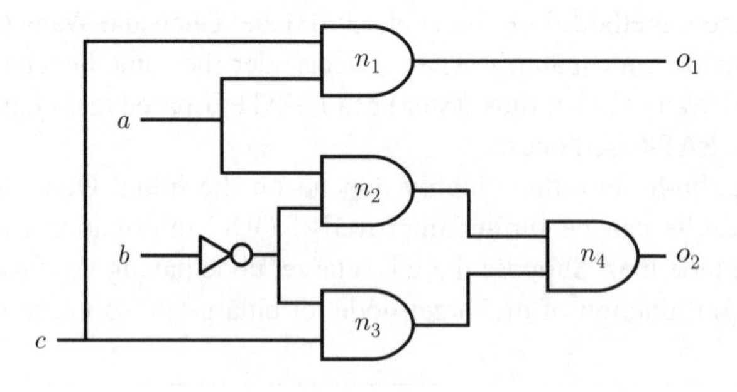

Figure 5.1 Example of ODCs and node merging

the output value of node n_3 is a don't care under such situation. Thus, replacing n_3 with n_1 does not change the overall functionality.

We can infer from Equation 5.1 that the larger the set of ODCs, the easier it is to replace the target node. This is because having more ODCs allows more logic difference between the target and the substitute nodes.

Computing exact ODCs explicitly is very time consuming. Hence, most algorithms use approximate ODCs instead. For example, ATPG techniques approximate ODCs by exploring compatible logic difference that will not be observed under any test patterns.

5.1.1.2 Node Merging

Rewriting Mishchenko et al. (2006) and sweeping Kuehlmann (2004) are the earliest node-merging algorithms known. They are effective and can be easily scaled for large circuits. But the constraint of these algorithms is that two nodes must be logically equivalent to be merged. This requirement is too strict and prevents us from finding more merging solutions.

In Plaza et al. (2007), the authors proposed a Boolean satisfiability-based (SAT-based) node-merging algorithm that computes the approximate ODCs of each node and applies SAT to check the correctness.

NAR (Chen and Wang 2009) is an ATPG-based algorithm that replaces a target node by another existing node inside the netlist. In Chen and Wang (2010), NAR was improved to allow replacement of a target node by an additional new node when no suitable existing nodes could be found for substitution.

With regard to speed, Chen and Wang (2010) runs faster because ATPG-based techniques are intrinsically more efficient than SAT-based techniques.

Mandatory assignments (MAs) are the values that have to be assigned to some nodes so as to activate and propagate the specific stuck-at fault. The NAR algorithm assumes that there are stuck-at faults ($stuck\_at\_0$ and $stuck\_at\_1$) occurring at the target node. It then calculates the MAs for each of the $stuck\_at\_0$ fault (s_a_0) and $stuck\_at\_1$ fault (s_a_1), which are denoted by MA_0 and MA_1, respectively. A node having different values in MA_0 and MA_1 and is not in the transitive fanout cone of the target node, whether or not it is an existing node in the circuit or an additional new node, is considered as a qualified substitute node for the target node.

5.1.2 Our Methodology ("Long tail" vs "Bump tail" Curves)

As discussed at the beginning, traditional node-merging techniques have some limitations in area reduction. In our methodology, we coupled it with a wire-based logic synthesis tool, namely rewiring, to help relaxing the limitation. Before describing the details of our algorithm, we first explain an underlying optimization methodology that helps us to explore better problem solutions.

Inspired by the orthogonal greedy coupling methodology Wu and Marek-Sadowska (1995), we designed two mutual-orthogonal area reduction strategies, a node-based one and a wire-based one. A certain strategy is applied trying to reduce the area, and a coupled strategy will then be deployed if the current strategy failed.

5.1.2.1 Orthogonal Greedy Coupling (OGC) Methodology

Two algorithms are considered mutual-orthogonal if they have "similar style and soundness" while with certain strategic "contrast" (e.g., route shortest net first vs route longest net first, or do single-seed grow routing vs multi-seed grow routing, as shown in Wu and Marek-Sadowska (1995)).

The authors observed the "greedy decaying effect" that happens when applying a deterministic algorithm to a complicated (NP-hard) problem and proposed the captioned OGC methodology to tackle this undesired phenomenon.

5.1.2.2 Greedy Decaying Effect (Long Tail Curve)

A greedy decaying effect can be described as an apparent effectiveness degrading after certain running point when applying a deterministic algorithm to a complicated (NP-hard) problem, a phenomenon observed in a field-programmable gate array (FPGA) routing experiment, where each routing tract forms an independent routing domain (the x-coordinate) and the y-axis gives the percentage of routing segments of the domain that are committed (used) in the final routing. We can observe that the usage rate of the routing segments for each tract domain drops drastically after the 10th routing tract, as shown in Figure 5.2, which is directly cited from the original paper.

5.1.2.3 Greedy Coupling Effect (Bump Tail Curve)

To eliminate the greedy decaying effect, the orthogonal greedy coupling idea was proposed, where two coupled routing algorithms "GG_S" (single-seed grow routing) and "GG_M" (multiple-seed grow routing), which is similar to Prime's versus Kruskal's of MST, are experimented. As shown in Figure 5.3, when a "coupling algorithm" is switched in, a sudden "performance jump" can be observed and the job completion can be achieved earlier than applying any single algorithm alone for the whole run. For a closer look, Figure 5.3(a) shows the result obtained by running a single algorithm all the way through the routing process. In the orthogonal greedy coupling flow, the algorithm is switched from "GG_S" to "GG_M" from the eighth iteration and the result is shown in Figure 5.3(b). We can see that the routing

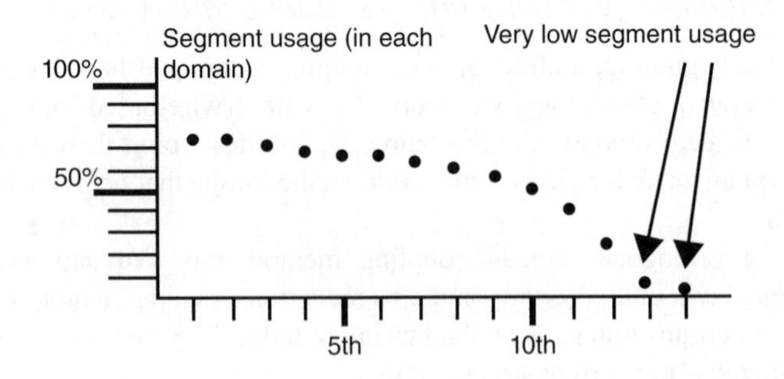

Figure 5.2 Typical greedy decaying effect in bin-packing router (Wu and Marek-Sadowska 1995)

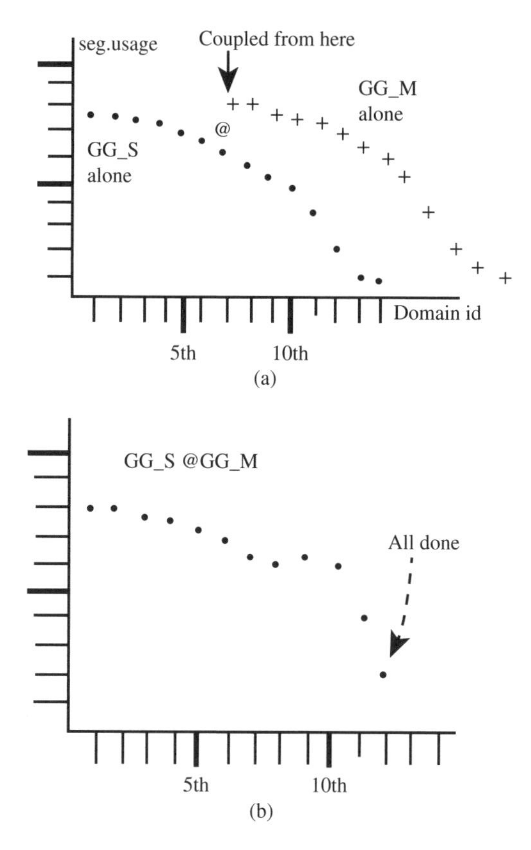

Figure 5.3 Example of orthogonal greedy coupling (Wu and Marek-Sadowska 1995). (a) Algorithm alone; (b) algorithm coupled

job can be completed earlier and two tracks are reduced compared to when running any algorithm alone.

Depending on the problem, algorithms can be coupled in various ways: for example, one algorithm runs first and its coupling algorithm takes over when the first gets depleted in its power. Intuitively, we may conjecture in this way: perhaps a different routing algorithm tends to "deplete" its optimization room or resource in a certain different "dimension" due to its intrinsic "deterministicness"; thus switching to another algorithm similarly sound but optimizing in a "contrast" direction might be easier to jump out of the trapping in a certain local minimum. Experimental data seem to justify this empirical conjecture. (Perhaps, in a Major League analogy, learned from a long-time playing experience, a baseball team always uses different starter and closer pitchers for better complement on different game stages to optimize its final winning chance.)

Our OGC approach is basically generalized from the above idea. In our flow, two coupling algorithms (node-based and wire-based) are coupled for a same area reduction goal. They are used to perform a coupling run in which the node-merging algorithm runs first and its coupling algorithm, rewiring, takes over for an "orthogonal" perturbance to explore a new performance jump.

5.1.2.4 Rewiring for Further Optimization

Though node merging is a well-known technique for area reduction, rewiring can perform the same function too. In fact, some earlier works (Chang and Marek-Sadowska 1996) already applied rewiring technique to simplify netlists.

In practice, node merging can be regarded as a special kind of rewiring technique. It removes several target wires (TWs) (all fanout wires of the target node) simultaneously, and has a constraint that all alternative wires must share the same source node (the substitute node).

Unlike node merging, which removes all fanout wires simultaneously, rewiring replaces the fanout wires one by one. The target node can be removed from the netlist too, when all its fanout wires are successfully removed. In terms of node removal, rewiring is not as efficient as node merging, as we cannot determine whether the target node can be successfully removed until one of its fanout wires is proven to be unremovable. But in some cases, rewiring can find a solution where node merging fails. In fact, rewiring can also be viewed as a special and extended node-merging technique that replaces the target node with multiple existing nodes. Figure 5.4 shows the difference between node merging and rewiring when removing the target node.

5.1.2.5 Rewiring for ODCs Shifting

From the previous section we know that the degree of difficulty of replacing a node is directly related to its ODCs. Intuitively, it is easy (difficult) to replace a node when the don't cares

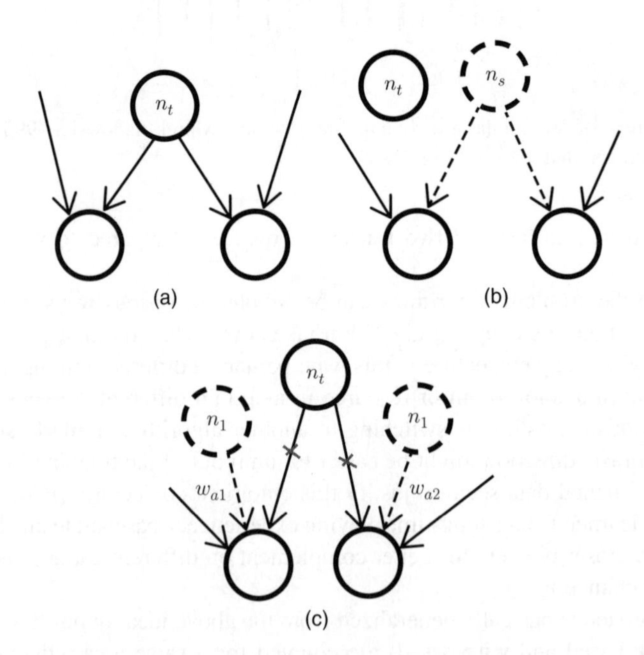

Figure 5.4 Difference between node merging and rewiring. (a) Original netlist; (b) removing n_t with n_s by node merging; (c) removing node n_t by rewiring

set is large (small). Though recent node-merging techniques can utilize the flexibility of ODCs, they cannot change the ODCs of nodes intentionally. In the following text, we will explain how rewiring can shift ODCs on the netlist to lead to a better structure for subsequent optimization.

Intuitively, a node closer to PIs, or more distant from POs, is likely to have a larger set of ODCs, as it tends to be less observable at the POs. Typically, if a gate has a fanout, that is, a primary output, the gate is fully observable and has no ODCs.

Therefore, during the rewiring process, we can intentionally push some critical nodes to be closer to PIs or further away from POs to increase their ODCs. We have the following heuristics.

Heuristic 1: Suppose node n_a is a PI or a node very close to a PI. Adding a new wire from n_a to another node n_b can reduce the distance from n_b to PIs, and consequently the ODCs of n_b and the nodes in its transitive fanin cone are increased.

Heuristic 2: Suppose $n_a \to n_b$ is an existing wire, and n_b is very close to POs. If we can disconnect n_a from n_b by replacing $n_a \to n_b$ with an alternative wire, the distance of n_a to POs may be increased. In that case, the ODCs of n_a and nodes in its fanin cone may be increased.

Typically, if a gate has more fanouts, it is more easily observed. This is because there are more paths to the primary outputs. As a result, the size of the ODCs associated with this node decreases, and the same applies to the nodes in its fanin cone. However, it is not always true because multiple fault effects may cancel each other due to the reconvergence of the node's fanouts. We have the following heuristic.

Heuristic 3: Suppose $n_a \to n_b$ is a fanout wire of n_a, whose fanout cone does not intersect the fanout cones of n_a's other fanouts. Then, deleting $n_a \to n_b$ increases the ODCs of n_a and the nodes in its fanin cone. The purpose of the restriction on the fanout cone of wire $n_a \to n_b$ is to ensure that there is no fanout reconvergence.

In the experiments in Chen and Wang (2010), every benchmark was processed with six iterations of optimization. The authors showed that even by running the optimization flow with the traditional node-merging algorithm NAR for five more iterations, the average improvement on area reduction could increase only 0.9% (from 5% to 5.9%). This implies that node-merging algorithms reach a local minimum rapidly after the first iteration. As discussed earlier, rewiring techniques can escape the local minimum by introducing perturbations on the netlist.

The perturbations introduced by rewiring can alter the circuit structure by replacing a wire with another wire without increasing the circuit size. Applying such unintrusive rewiring transformations repeatedly by proper guidance can lead to a better circuit structure for subsequent optimization processes. The heuristics guiding the transformation process will be discussed in the next section.

Figure 5.5(a) and (b) gives a small example to show how rewiring and node merging are applied together to achieve a better optimization result. Originally, no merge solution can be found in Figure 5.5(a). But if we can remove wire $n_2 \to n_4$, or replace it with an alternative wire, then node n_1, which is in the fanin cone of n_2, can be replaced by n_6. Figure 5.5(b) shows the optimized netlist.

Rewiring has been proven to be efficient. In Yang et al. (2010), the authors demonstrated that the complexity of replacing a wire by ECR is very close to merging a node and that the rewiring rate (the rate of successfully replacing a target wire with its alternative wires) is nearly 40%. The number of nodes in the circuit can be kept from increasing during the rewiring process.

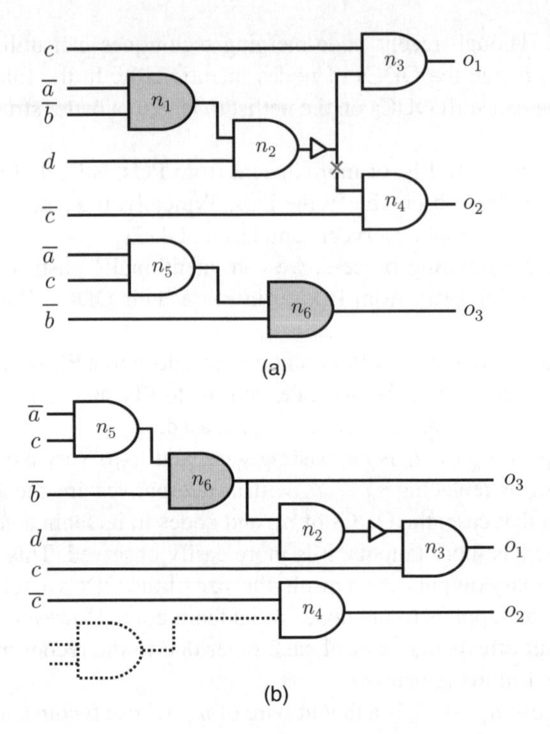

Figure 5.5 Example of coupling rewiring and node merging. (a) Rewiring for ODCs shifting; (b) netlist after rewiring and node merging

5.1.3 Details of our Approach

In this section, we will describe our new approach for area reduction in detail. We will focus on how the node-merging technique and rewiring technique are coupled.

In an iteration, our approach mainly consists of two stages: the optimization stage and the perturbation stage. In the optimization stage, we try to reduce the number of nodes by both node merging and rewiring techniques. In the perturbation stage, rewiring is applied to perturb the netlist so as to escape from a local minimum.

5.1.3.1 Optimization Stage

In Plaza et al. (2007) and Chen and Wang (2010), only the node-merging technique is applied to reduce the number of nodes in the netlist. However, in some cases rewiring can find a solution to remove the target node while node merging fails.

In our approach, for each node, we first try to replace it with an existing node or an additional new node by node merging. If that fails, rewiring is then performed on the fanout wires of the target node one by one. If all fanout wires can be successfully removed, the target node can be removed from the netlist. If we cannot remove a target node, reducing its number of fanouts will likely have the effect of increasing the ODCs of the nodes in the target node's fanin cone, as well as the ODCs of the target node.

Before rewiring, we assign a priority to every fanout wire of the target node according to its potential for ODC optimization. A wire with higher priority will be treated earlier. According to the discussion in the previous section, the fanout wires with following features will be assigned with higher priority:

1. Wires whose fanout cones do not intersect the fanout cones of other fanout wires. According to Heuristic 3, deleting such wires can increase the ODCs of the target node.
2. Wires that are close to POs. Deleting them can probably push the target node further away from POs according to Heuristic 2.

During rewiring, according to Heuristic 1, the target wire is replaced by the alternative wire that is the closest to PIs in our scheme.

As discussed previously, performing rewiring on the target node can help in increasing the ODCs of the target node and also the ODCs of all the nodes in the target node's fanin cone. Obviously, a node nearer to the POs has a larger fanin cone. Hence, target nodes are selected in the reverse topological order from POs to PIs in the optimization flow so that more nodes can potentially be benefited from subsequent optimization.

Algorithm 5.1 describes the flow of the optimization stage. In practice, we choose NAR as the node-merging technique and ECR as the rewiring engine, as they are both efficient and powerful. The parameter max_fanout_limit is used to decide whether rewiring should be performed. If the number of fanout wires of a node exceeds max_fanout_limit, rewiring is not performed because it is practically hard to optimize such nodes.

Algorithm 5.1: Optimization algorithm

```
1  begin
2      C' ← C;
3      for each node n_t in C in DFS order from POs to PIs do
4          N_s ← Find substitute node by node merging algorithm NAR ;
5          if success then
6              C' ← Replace n_t with a substitute node n_s ∈ N_s That is the closest to PIs;
7          else if fanout number of n_t < max_fanout_limit then
8              Sort fanout wire by priority;
9              for each fanout wire w_t do
10                 W_a ← Find alternative wires by rewiring engine ECR;
11                 if success then
12                     C' ← Replace w_t with an alternative wire;
13                     w_a ∈ W_a whose source node is the closet to PIs;
14     return C';
15 end
```

5.1.3.2 Perturbation Stage

As the local optimum will be reached gradually, we perform a perturbation stage between two optimization stages. We extend the method in Chang and Marek-Sadowska (1996) as follows: We try to transform the netlist into the structure suitable for optimization by performing rewiring repeatedly. Unlike the optimization stage, target nodes are selected in the topological order from PIs to POs in the perturbation stage. For each target node, we treat each of its fanout wire as the target wire for rewiring. Since estimating the change of ODC is very time consuming, some heuristics are adopted to choose the best alternative wire to be added for a target wire. For a target wire $n_a \rightarrow n_b$, we follow the rules derived from Heuristic 1 to choose the best alternative wire. Suppose $n_c \rightarrow n_d$ is a candidate alternative wire. In order to optimize the ODCs, the destination node of the alternative wire n_d has to be a dominator of the target wire, and the source node of the alternative wire n_c must be a PI or closer to a PI than the source node of the target wire n_a.

In Figure 5.6, Cone I contains n_a and its fanin cone. After a fanout wire of n_a is removed, we assume that the ODCs of the nodes in Cone I are increased. Cone II contains n_c and its fanin cone. If an additional fanout wire is added to n_c, we consider that the ODCs of the nodes in Cone II are decreased. Cone III contains all the nodes in the intersection of the fanin cone of n_d and the fanout cone of n_b. As demonstrated in Chang and Marek-Sadowska (1996), the ODCs of the nodes in this area will be increased after the transformation.

The circuit will be transformed with an alternative wire $n_c \rightarrow n_d$ if the following equation is satisfied:

$$Size(ConeI) + Size(ConeII) > Size(ConeIII) \tag{5.2}$$

Equation 5.2 implies that we accept the transformation if the number of nodes that have their ODCs increased is larger than the number of nodes that have their ODCs decreased.

Algorithm 5.2 describes the flow of the perturbation stage. As in the optimization stage, error cancellation rewiring (ECR) is adopted as the rewiring engine.

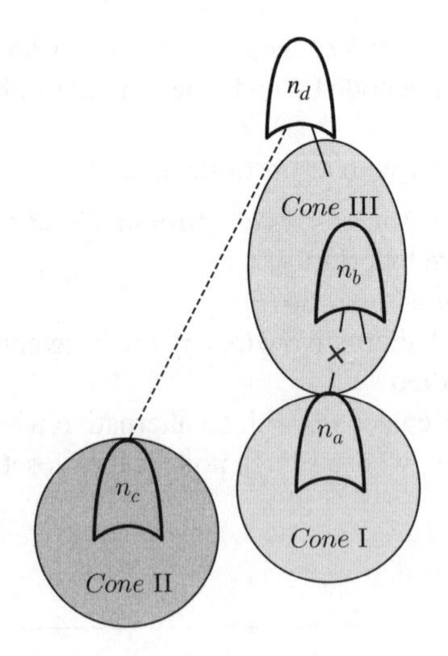

Figure 5.6 Rewiring for perturbation

Algorithm 5.2: Perturbation algorithm

```
1  begin
2      C' ← C;
3      for each node n_t in C in topological order from POs to PIs do
4          for each fanout wire w_t do
5              W_a ← Find alternative wires by rewiring engine ECR;
6              Sort alternative wires W_a in the order of the distance between their source
               node and PIs;
7              for each alternative wire w_a ∈ W_a do
8                  Calculate Size(ConeI), Size(ConeII), Size(ConeIII);
9                  if Size(ConeI) + Size(ConeII) > Size(ConeIII) then
10                     C' ← Replace w_t with an alternative wire;
11                     break;
12     return C';
13 end
```

5.1.3.3 The Main Flow

In our flow, considering the tradeoff between performance and runtime, we choose to run the optimization stage three times and a perturbation stage between two runs of the optimization stage.

5.1.4 Experimental Results

We implemented our algorithm in C++. Our algorithms were tested against the IWLS 2005 benchmark suite, which was also used by Plaza et al. (2007) and Chen and Wang (2010). The tests were run on a 2.8 GHz 1 GB RAM Linux OS system. We only consider the combinational parts of the benchmarks. For a fair comparison, we synthesized each benchmark into an and-inverter graph (AIG) with the *resyn2* script using the ABC (Berkeley Logic Synthesis and Verification Group) package (which performs local circuit rewriting optimization) and count the number of and-inverter nodes before and after optimization. The same procedure was also used in Plaza et al. (2007) and Chen and Wang (2010). The correctness of our optimization method was verified by the equivalence checking tool in the ABC package.

Table 5.1 summarizes the experimental results. As we could not obtain five of the benchmarks used in Plaza et al. (2007) and Chen and Wang (2010), we only tested the 18 available cases. For most cases, the initial synthesized circuit sizes are comparable to those reported in Chen and Wang (2010). The difference may be due to the difference of the ABC version (we used the latest version of ABC in our work). Two cases have more than 30% difference.

The benchmarks are listed in the first column of the table. Columns N and Nr list the number of and-inverter nodes before and after the optimization, respectively. Column % shows the percentage of circuit size reduction. And column $T(s)$ shows the CPU time. The statistics of our approach, NAR (Chen and Wang 2010), and the SAT-based node-merging approach (Plaza et al. 2007) are presented side by side. Since the authors of Plaza et al. (2007) did not

Table 5.1 Experimental results of area reduction by using approaches in Plaza et al. (2007) and Chen and Wang (2010) and our approach

Benchmark	Our approach				Chen and Wang (2010)				Plaza et al. (2007)	
	N	Nr	%	$T(s)$	N	Nr	%	$T(s)$	%	$T(s)$
s9234[a]	809	776	4.1	12	1,353	1,323	2.2	0.2	1.2	8
pci_spoci_ctrl	824	591	28.3	221	878	757	13.8	0.4	9.2	6
s13207[a]	848	803	5.3	14	2,108	2,043	3.1	0.8	1.8	17
i2c	935	839	10.3	66	941	894	5	0.2	3.2	3
systemcdes	2,442	2,178	10.8	227	2,641	2,580	2.3	1.2	4.7	9
spi	3,208	3,045	5.1	415	3,429	3,383	1.3	5.6	1.3	84
tv80	7,190	6,445	10.4	7,191	7,233	6,813	5.8	20.3	7.1	1,445
s38584	7,331	7,156	2.4	1,695	9,990	9,836	1.5	15.1	0.8	223
s38417	8,045	7,891	2	620	8,185	8,105	1	1.5	1	275
mem_ctrl	8,578	6,756	21.2	1,914	8,815	7,287	17.3	13.8	18	738
systemcaes	9,930	9,664	2.7	3,693	10,585	10,386	1.9	30.7	3.8	360
ac97_ctrl	10,287	10,185	1	728	10,395	10,364	0.3	3.1	2	188
usb_funct	13,339	12,537	6	2,508	13,320	12,868	3.4	11.4	1.4	681
pci_bridge32	16,423	16,163	1.6	3,589	17,814	17,599	1.2	19.7	0.1	1,134
aes_core	20,472	19,412	5.2	5,282	20,509	20,195	1.5	22.7	8.6	1,620
b17	33,219	30,114	9.3	9,096	34,523	33,204	3.8	205.5	1.6	5,000
wb_conmax	41,230	36,293	12	8,101	41,070	38,880	5.3	48.4	6.2	5,000
des_perf	74,546	67,600	9.3	9,372	71,327	69,421	2.7	84.7	3.7	5,000
Average			**8.2**				**4**		**4.2**	
Total				54,744				485.3		21,796

[a]Benchmarks with large different size with Chen and Wang (2010) after preoptimization.

report the node number after optimization, we only list the node reduction percentage and runtime. Because of the version difference of ABC for pre-optimization, the original number of nodes N is slightly different from that of Chen and Wang (2010).

The experiments show that by coupling the rewiring and node-merging techniques, the results can yield nearly double the improvement compared to when applying the node-merging algorithm alone. We believe that the improvement is mainly due to the effective ODC shifting algorithms proposed by this new scheme.

In our flow, we implemented our own version of the NAR and ECR algorithms. As we built the algorithms atop a complicated integrated database supporting operations on both logical and physical electronic design automation (EDA) operations, in a similar single NAR run our NAR takes 10 times the CPU time than the original implementation in Chen and Wang (2010). The CPU time spent in our flow is about 10 times that of one NAR run, which is due to the three iterations of optimizations with a perturbation stage inserted in between. Besides, rewiring is intrinsically more complicated than node merging. There are three iterations in our approach. In Table 5.2, we list the optimization details of each iteration. The statistics shows that our approach already outperformed the previous node-merging algorithm in iteration 1. The advantage of our approach over the previous approaches is that the results can still be improved after a few iterations for perhaps an effective circuit perturbation capability. Nearly 3% further improvement could be achieved after two more iterations.

Table 5.2 Iterations of optimization in our approach

Benchmark	N	iteration 1		iteration 2		iteration 3	
		Nr	%	Nr	%	Nr	%
s9234	809	791	2.2	786	2.8	776	4.1
pci_spoci_ctrl	824	657	20.3	619	24.9	591	28.3
s13207	848	808	4.7	805	5.1	803	5.3
i2c	935	875	6.4	856	8.4	839	10.3
systemcdes	2,442	2,259	7.5	2,207	9.6	2,178	10.8
spi	3,208	3,156	1.6	3,118	2.8	3,045	5.1
tv80	7,190	6,644	7.6	6,473	10	6,445	10.4
s38584	7,331	7,190	1.9	7,160	2.3	7,156	2.4
s38417	8,045	7,937	1.3	7,909	1.7	7,891	2
mem_ctrl	8,579	6,889	19.7	6,787	20.9	6,756	21.2
systemcaes	9,930	9,712	2.2	9,698	2.4	9,664	2.7
ac97_ctrl	10,287	10,230	0.6	10,193	1	10,185	1
usb_funct	13,339	12,679	4.9	12,561	5.8	12,537	6
pci_bridge32	16,423	16,218	1.2	16,174	1.5	16,163	1.6
aes_core	20,472	19,766	3.4	19,568	4.4	19,412	5.2
b17	33,219	31,459	5.3	30,432	8.4	30,114	9.3
wb_conmax	41,230	38,385	6.9	36,576	11.3	36,293	12
des_perf	74,546	71,471	4.1	68,574	8	67,600	9.3
Average			5.6		7.3		8.2

5.2 Postplacement Optimization

In spite of having been studied for over 40 years, placement still remains one of the most active EDA research areas. Placement is such a fundamental EDA step that it can determine the overall quality of a chip's final layout in terms of area and performance. In today's known EDA flows, the placement procedure is called after the completion of logic synthesis, and the inputs of most placement tools do not include logic view information. However, with the rewiring technique, we can insert logic restructure/perturbation into the placement stage by rewiring, and this process can give us another solution space to further optimize the placement, which cannot be done by any other traditional physical optimization techniques.

5.2.1 Wire-Length-Driven Rewiring-Based Postplacement Optimization

In this section, we introduce how to use rewiring technique to optimize the wire length on an already fully optimized placement. Since there are various of metrics to calculate the wire length, we just use the half parameter wire length (HPWL), which is nearly the most common metric during the placement stage within our algorithm.

Figure 5.7 illustrates an obvious example showing a potential HPWL reduction obtainable purely by logic restructuring with the placement intact. In Figure 5.7(b), the HPWL of the original net driven by $g1$ is 180 units. We observe that by logic restructuring, we can add a connection from $g3$ to $g7$ and replace the connection from $g1$ to $g5$ without changing the logical functionality of the whole circuit. As under the current placement, the HPWL of the

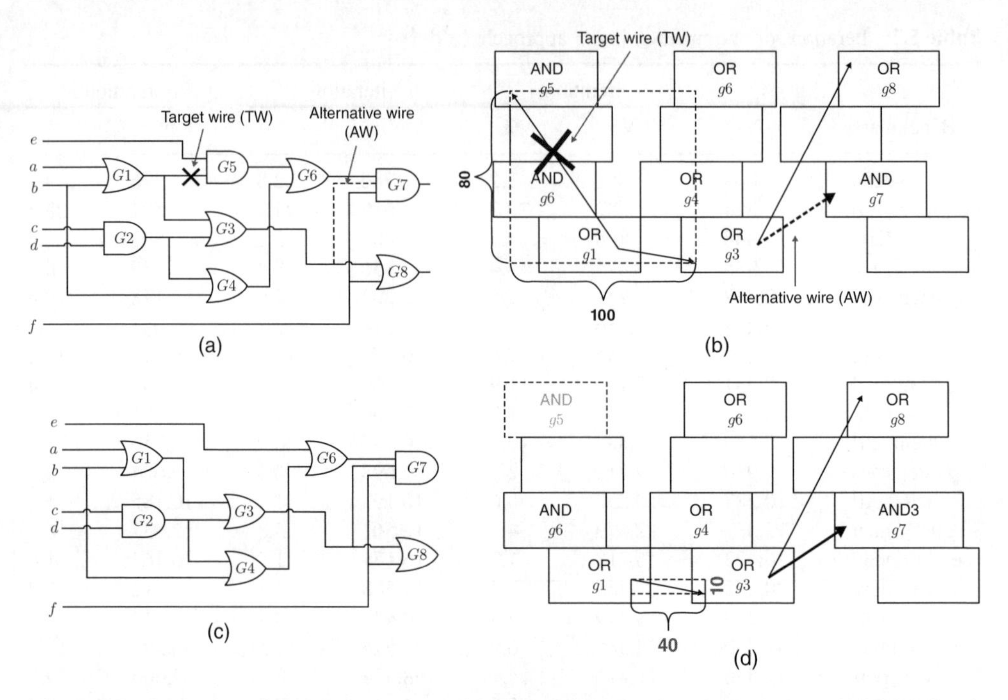

Figure 5.7 Reduction of HPWL by logic rewiring. (a) Initial circuit in logical view; (b) initial circuit in physical view; (c) optimized circuit in logical view; (d) optimized circuit in physical view

net driven by $g1$ is reduced to 50 units while the HPWL of net driven by $g3$ does not increase, the total HPWL is reduced by 130 units after applying this (rewiring) logic restructuring.

As logic rewiring creates alternative wires to replace existing target wires, the basic idea is to use logic rewiring to iteratively remove each **troublesome** target wire by adding its alternative wire to yield a new placement with a less total HPWL. In this section, a "troublesome" wire is one that, when removed, largely reduces the total HPWL.

5.2.1.1 Target Wire Recognition

Our algorithm examines each target wire (TW) and calculates the HPWL reduction when it is removed. So, we temporarily remove the wire and estimate the change in HPWL. If the HPWL change is negative, we confirm the wire removal and pass the target wire to the rewiring engine. The HPWL estimation after removing the target wire is shown in Figures 5.8 and 5.9.

Figure 5.8 shows an example of estimating the HPWL reduction when a single fanout net is removed. The estimated HPWL reduction consists of three parts. An obvious one is the reduction due to disconnecting the target wire, which is shown by the center filled area in Figure 5.8(b). As the source gate has only one fanout, the gate will become redundant and the fanin cone of the source can also be removed without affecting the overall logic functionality. The estimated HPWL reduction due to disconnecting the fanin cone of the source gate is shown by the left filled area in Figure 5.8(b). After the TW is removed, the sink gate of the TW can be moved to a better location to reduce the HPWL of its fanin and fanout nets. It has been shown that the local HPWL is minimal if the gate is placed at the x-median and y-median of

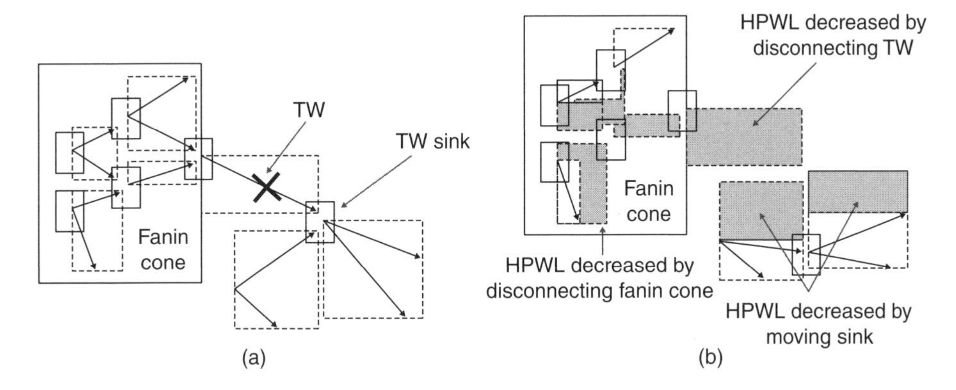

Figure 5.8 Estimation of HPWL of a single fanout net. (a) Before removing TW; (b) after removing TW

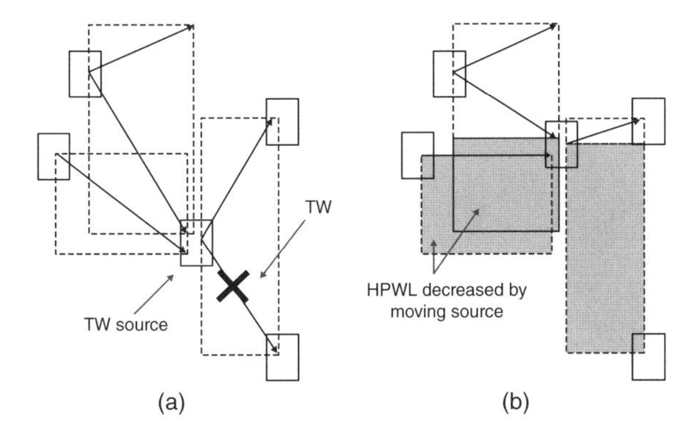

Figure 5.9 Estimation of HPWL of multiple fanout nets. (a) Before removing TW; (b) after removing TW

all its source pins of the fanin nets and sink pins of its fanout nets. Such an HPWL reduction is shown by the right filled area in Figure 5.8(b).

Figure 5.9 shows the steps to estimate HPWL reduction when TW is in an n-pin net where $n \geq 2$. The TW source gate cannot be removed after removing the TW, so WRIP cannot simply disconnect fanin cone. In this case, the potential HPWL reduction of disconnecting the fanin cone will always be 0. But we can relocate TW source gate to achieve further HPWL reduction. The change in HPWL by moving TW source is shown by the left filled area in Figure 5.9(b).

5.2.1.2 Alternative Wire Selection

When we attempt to remove a target wire, the rewiring engine usually returns more than one AW, each of which can replace the target wire. For each AW candidate, we temporarily add it into the netlist and estimate the HPWL increase. More importantly, we need to *recalculate* the HPWL reduction due to AW addition because TW and AW may share the same source or sink gates. After testing all AW candidates, the best candidate that gains the most HPWL reduction will be adopted.

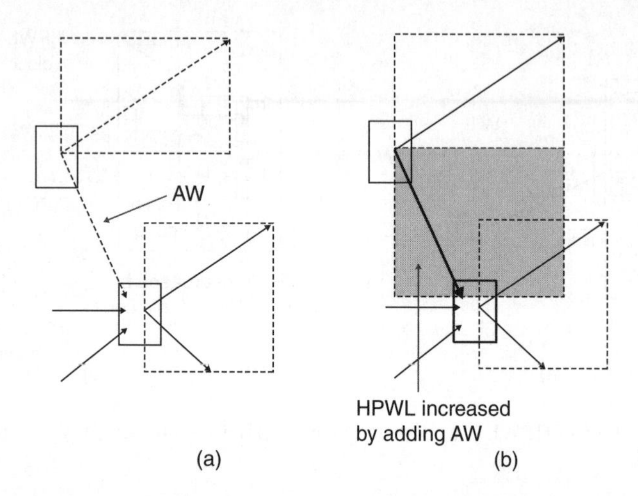

Figure 5.10 HPWL increase due to AW addition. (a) Before adding AW; (b) after adding AW

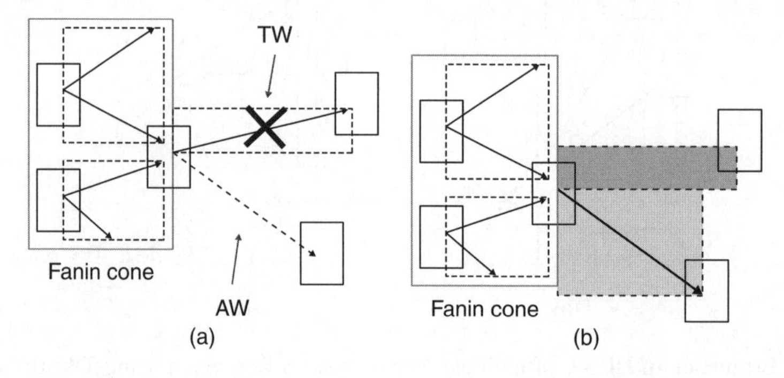

Figure 5.11 Revised HPWL estimation when TW and AW have the same source. (a) Before adding AW; (b) after adding AW

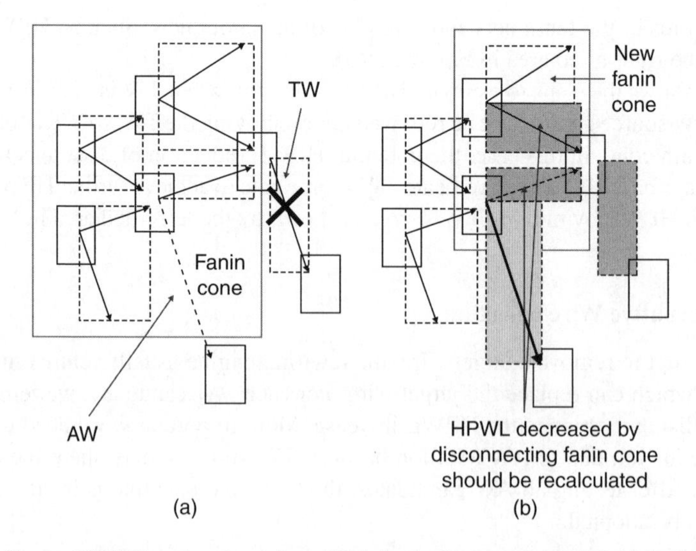

Figure 5.12 Revised HPWL estimation when AW is in the fanin cone of TW. (a) Before adding AW; (b) after adding AW

Figure 5.10 shows the most common case when adding the AW to the circuit. The highlighted area shown in Figure 5.10(b) is the HPWL increase due to the AW addition.

Figure 5.11 shows a special case where TW and AW have the same source gate. In this case, as the source gate has only one fanout that is removed, the fanin cone of the source could have been disconnected before adding the AW and we could have HPWL reduction by disconnecting the fanin cone. After adding the AW, the previously mentioned source gate cannot be removed because it has to drive the AW. HPWL reduction by disconnecting fanin cone has to be revised to 0 because no fanin is able to be disconnected, as shown in Figure 5.11(b).

Figure 5.12 shows another special situation where the source gate of AW belongs to the fanin cone of TW. In this case, the fanin cone will change after creating AW, and the HPWL reduction by disconnecting fanin cone has to be recalculated. The revised estimation of HPWL reduction is shown in Figure 5.12(b).

The total flow of our algorithm is shown in Algorithm 5.3. We traverse each wire and identify "troublesome" target wires if the TW removal can potentially reduce HPWL. We call rewiring engine with the target wire, and it generates a list of alternative wire candidates. For each TW,

Algorithm 5.3: Wire-length-driven rewiring-based postplacement optimization flow

```
1  begin
2      for each target wire TW = (i, j) do
3          temporarily remove TW and estimate HPWL reduction(ΔHPWL) by the
           following 4 lines;
4          estimate ΔHPWL by disconnecting TW;
5          estimate ΔHPWL by moving TW sink i;
6          estimate ΔHPWL by moving TW source j;
7          estimate ΔHPWL by disconnecting fanin cone of i;
8          calculate total HPWL reduction after removing TW (ΔHPWL_TW);
9          if ΔHPWL_TW ≤ 0 then
10             continue;
11         call rewiring engine and collect all AW candidates for TW;
12         ΔHPWL_best = 0;
13         AW_best = NULL;
14         for each AW candidate do
15             temporarily add AW into the netlist;
16             revise ΔHPWL by disconnecting fanin cone of i;
17             estimate HPWL increased by adding AW;
18             calculate total HPWL reduction after adding AW (ΔHPWL_AW);
19             if ΔHPWL_best < ΔHPWL_AW then
20                 ΔHPWL_best = ΔHPWL_AW;
21                 AW_best = AW_current;
22         if AW_best ≠ NULL then
23             permanently remove TW and add AW_best into circuit;
24 end
```

WRIP estimates the HPWL reduction for each TW and AW pair and chooses the best pair that achieves the best HPWL reduction.

5.2.1.3 Experimental Results

We implement the algorithm in C++ and integrate it into an industrial EDA tool. All the experiments are performed on a Linux workstation with Intel(R) Xeon(R) CPUs X7350 running at 2.93 GHz. We tested WRIP on seven real, high-performance ASIC designs.

Table 5.3 shows the basic information of the benchmarks with 100,000 cells to 1 million cells. All benchmarks are real commercial circuit designs. In Table 5.3, column 2 shows the HPWL at the end of an industrial EDA placement tool; column 3 is the number of cells, which includes sequential cells such as registers and latches, arithmetic cells such as adders and subtractors, and simple combinational cells such as NAND, MUX, AOI gates, and so on; column 4 is the number of nets and column 5 is the number of wires, which implies that each circuit net has 3 fanouts on average.

Table 5.4 demonstrates the accuracy of our estimation of HPWL reduction. We compare the estimated HPWL reduction with the real HPWL reduction obtained after the diffusion-based

Table 5.3 Benchmark information

Circuit	HPWL	#Cell	#Net	#Wire
ibm1	17,307,504	98,631	100,315	316,350
ibm2	50,933,622	318,703	324,102	846,651
ibm3	55,531,283	292,141	295,995	823,244
ibm4	52,551,332	295,781	300,312	826,153
ibm5	48,316,683	322,038	327,605	865,159
ibm6	156,464,726	394,915	408,776	1,340,470
ibm7	383,720,081	1,214,478	1,302,436	4,125,914
Total	764,825,231	2,936,687	3,059,541	9,143,941

Table 5.4 Real versus estimated HPWL reduction

Circuit	Real HPWL reduction	Estimated HPWL reduction	Difference	Difference ratio (%)
ibm1	263149	262519	630	0.24
ibm2	494169	532313	38144	7.72
ibm3	963262	1235655	272392	28.28
ibm4	1138092	1519884	381792	33.55
ibm5	889164	970413	81249	9.14
ibm6	2910768	2959316	48548	1.64
ibm7	5298495	5471992	173497	3.27
Total	11957099	12952092	996252	7.69

Table 5.5 Effectiveness of our framework for wirelength reduction of placement and global routing

Circuit	Placement				Global routing		
	Initial HPWL	HPWL reduction	Ratio (%)	Runtime (s)	Initial Steiner WL	Steiner WL reduction	Ratio (%)
ibm1	17307504	337819	1.95	130	20487793	509694	2.43
ibm2	50933622	640299	1.26	352	65201994	1123606	1.69
ibm3	55531283	1210168	2.18	443	67858981	1956429	2.80
ibm4	52551332	1433974	2.73	490	66365291	2437595	3.54
ibm5	48316683	1137464	2.35	463	58014446	1856664	3.10
ibm6	156464726	3818605	2.44	1858	199685147	4114180	2.02
ibm7	383720081	6915837	1.80	4338	805460947	7337833	0.91
Total	764825231	15494166	2.02	8074	1283074599	19336001	1.51

legalization. The results show a mere 7.69% difference on comparing the estimated and real reduction, on average. Our algorithm involves gate creation, deletion, and movement, all of which may trigger Legalization, which would lead to a change in HPWL reduction. Compared to the final HPWL reduction shown in column 2, we can see that the estimated HPWL reduction shown in column 3 is slightly more optimistic for almost all circuits except case 1. Among our test cases, the cell density of "ibm3" and "ibm4" are much higher than other designs, so that legalization has a bigger impact to the change of HPWL reduction.

The column "placement" of Table 5.5 shows the HPWL reduction of our algorithm. Column 2 shows the initial HPWL, and columns 3 and 4 show the HPWL reduction obtained by our scheme. Column 5 shows the total runtime (for both rewiring engine and manager) for each design. The results show that we can reduce the total HPWL by 2.02% on average on top of a fully optimized industrial placement, which is a significant improvement on HPWL reduction on a fully optimized placement. The CPU time increase versus number of cells of circuits is nearly linear, which implies the effectiveness and efficiency of our algorithm.

We find that the HPWL reduction we gain can be well reflected in the subsequent physical design stage. We compare the wirelength between initial circuit and optimized circuit after performing global routing and use a more accurate Steiner wirelength model to calculate the total wirelength after global routing. The results are shown in column "global routing" of Table 5.5. We can see the Steiner wirelength can be reduced by 1.51% on average.

5.2.2 *Timing-Driven Rewiring-Based Postplacement Optimization*

The postplacement optimization is usually performed after the global placement and iterations of the optimization processes for wire length, timing, routability, and so on. Thus the postplacement process cannot perturb the placement much in case of losing the benefit gained by the previous optimization. Rewiring can be an ideal tool for these purposes since each rewiring process produces nearly the minimum changes (only one wire changed).

A key challenge of a postplacement timing-driven logic synthesis is the lack of an accountable principle for decision making based on a global picture of the whole circuit. In this section,

Figure 5.13 Figure Slack distribution graph of an industrial design

we design a heuristic, that is, able to make "natural" decisions based on the whole timing picture during the postplacement logic synthesis, which avoids many futile logic rewiring process and makes the flow efficient. For this purpose, we build a slack distribution graph to help us guide logic rewiring to shift the logic resource efficiently.

A slack distribution graph is a mapping from the locations of circuit cells to their slack values. An example of such a graph of an industrial customer design created by an industrial EDA tool is shown in Figure 5.13. The X-axis and Y-axis are the column index and row index of the chip, respectively. We can find that cells with negative slack values gather to form a big critical area, and the most critical areas (at the center of the figure) are surrounded by near-critical areas.

With the slack distribution graph built, one may be easily tempted to apply the rewiring process to improve on the most slack violated locations (wires with the most negative slack values) first (and directly). However, learning from our experiments, we find that it might not be the best strategy to remove the most critical wires directly. Logic rewiring (also similar to other logical restructuring techniques) is able to replace the target wires only by nearby alternative wires which may be within a few logic levels/hops to the target wires. Unfortunately, the critical timing paths often cover a large area and the longest paths are often surrounded by other nearby similarly critical paths (as shown in Figure 5.13) since the nearby signals are more likely to stem from the same already slack-violated source node. The original wire removed from the longest path and the new wire added into the near-critical path often result in the generation of a new longest path, which makes the wire exchange likely to be futile. And optimizing the

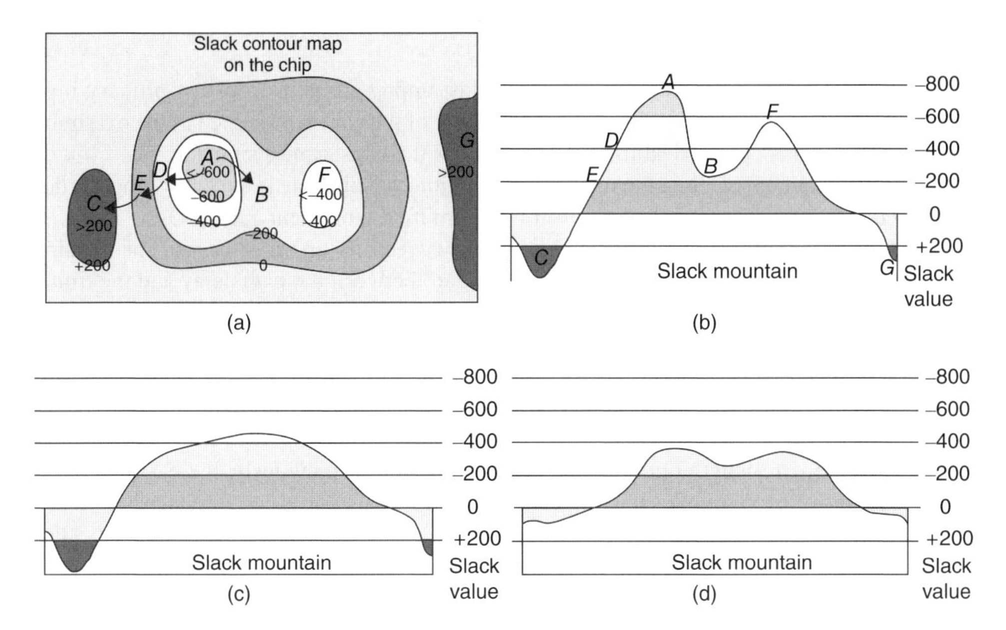

Figure 5.14 Optimizing circuit from slack mountain boundary resulting in more delay improvement. (a) Initial slack contour; (b) initial slack mountain; (c) optimizing slack mountain from peak first; (d) optimizing slack mountain from boundary first

circuit from the most critical path first is likely to be stuck at the local minimum solutions. This situation is exactly reflected by the natural mountain terrain on earth, where a higher mountain is always surrounded by other mountains of closer heights and, if we simply try to "blindly" shovel the earth from the highest peak and dump it at its nearby spots, we can be trapped in some futile cycling loop easily.

Therefore, we introduce a novel framework to optimize circuit delay by building a slack distribution graph to guide logic rewiring for the removal of critical wires in an effective order. Figure 5.14 illustrates a simple example explaining the idea of our scheme. Figure 5.14(a) is a slack distribution graph (slack contour map) of the initial circuit. We can model the circuit as a "slack mountain," as illustrated in Figure 5.14(b), with higher peaks (A and F areas) representing higher timing violating (negative) slacks and valley or sea (C and G areas) representing (positive) slack areas that can be used to receive logic shifted from violated areas (only the minimal or maximal slack values of each column are shown for clarity). If we start the shifting from peaks, we will not be able to know the best direction (toward deepest sea area C) to shift the logic resource, and the process is likely to rewire toward a local minimum trapping point B, as shown in Figure 5.14(c). However, in our scheme, if we shift the logic from the deepest sea or valley areas (like C and G) upward repeatedly (e.g., shifting logic from E to C and following with $D \rightarrow E$ and $A \rightarrow D$), at the end the deeper areas can be more fully utilized to receive logic shifted from the peak areas, and a more evenly distributed slack terrain can be produced, as shown in Figure 5.14(d). The example shows that our scheme can avoid being stuck at some local minimum solutions better and makes better decisions based on a global view.

5.2.2.1 Algorithm Flow

A circuit delay is defined as the largest signal delay among all paths from any primary input (PI) to any primary output (PO). The timing constraint of the circuit is defined as the maximum allowed circuit delay. The arrival time at a pin/gate is the latest time a signal arrives from the PIs inside its fanout cone, while the required arrival time at a pin/gate is the latest time the data is required to be present at it in order to fulfill the timing requirement at the POs.

The slack of a pin/gate is the difference (gap) between its arrival time and the required arrival time. The slack of a path is the difference (gap) between the path delay and the timing constraint. The slack of a wire (i, j) is the worst slack between i and j. Positive slacks can be considered as safety margins in circuit timing, while negative slacks represent timing violations. A gate is critical if it has a negative slack. The worst negative slack (WNS) of the circuit is the largest violations among all slacks in the circuit. Our goal is to improve the WNS as much as possible.

Similar to the wire-length optimization process introduced previously, the basic idea is to iteratively remove each troublesome target wire by adding its corresponding alternative wire to yield a new netlist with a better WNS. Intuitively, a wire is "troublesome" when its slack is worse than that of most other wires. Removing these troublesome wires greatly improves the WNS of the circuit. However, as we will show in this section, the order of wire processing significantly affects the quality of the result because timing optimization by rewiring can easily get stuck in a local minimum. Therefore, we propose the slack distribution graph to guide the wire processing order and update the graph during the iterations.

Given a circuit with its placement, we first compute the slack of each gate/wire based on the user-defined timing constraint. The most critical paths are identified, and a heuristic search is invoked to construct the slack mountain around the paths. The wires on the boundary of the slack mountain will first be processed before those near the peaks.

In an attempt to remove a troublesome wire, the alternative logic is identified in the area that is less critical based on the slack distribution graph. A set of alternative logic choices is found by the rewiring engine. An overlap-free incremental placement is performed after the logic restructuring to keep the placement legal. A fast estimation of the local slack change is performed for each alternative logic candidate to select the AW with the most estimated slack reduction. At the end of the loop for each TW, an accurate global static timing analysis (STA) is invoked to confirm the effectiveness of the selected AW in reducing WNS.

We can find that the key idea of our scheme is to figure out a local slack mountain around the most critical path and first shift the logic from boundaries of the slack mountain to avoid getting stuck at local minimum. The blocks (i.e., logic gates) of the slack mountain are collected by a heuristic search where blocks are expanded in the order of their slack values from worse to better. The flow is shown in Algorithm 5.4.

After the construction of slack mountain, we observe that the **negative** slack values of boundaries are usually close to zero, while those at the peaks are much worse. At a result, we can sort all wires in the slack mountain and process them one by one from the most non-critical (yet negative) to the most critical. Thus the slack mountains can be gradually shoveled down from boundaries toward the peaks.

When we attempt to remove a target wire, the rewiring engine usually suggests more than one AW to be chosen as the alternative logic. We use the slack distribution graph to guarantee that the original logic is shifted from a critical area to a less critical area. We first compute the gradient value $\vec{g_t} = (x_g, y_g)$ of TW in the slack distribution graph and the distance vector

Algorithm 5.4: Identifying local slack mountain

```
1 Input: Most critical path p, maximum size M of priority queue;
2 begin
3     Push all blocks of p into priority queue Q with their slack values as priority key;
4     while Q is not empty and Q is not full do
5         b = pop(Q);
6         for each fanin/fanout block fb of b do
7             Push fb into Q with its slack value as priority key;
8 end
```

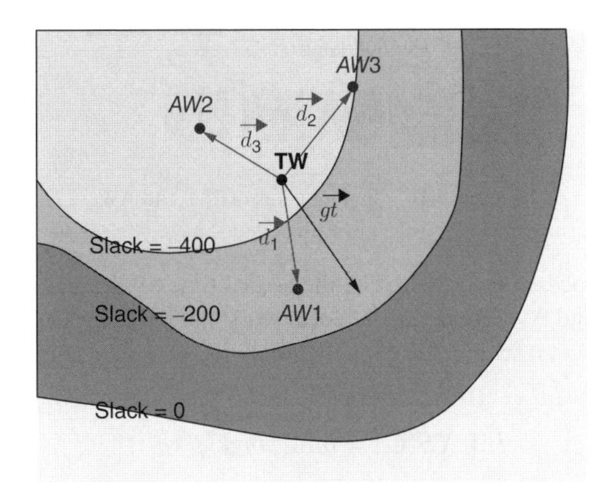

Figure 5.15 Example of AW selection

$\vec{d} = (dx, dy)$ between TW and AW. The AW will be taken as an available candidate of alternative logic if the following inequality holds:

$$
\begin{aligned}
slack_{aw} - slack_{tw} &> 0 \\
\vec{g_t} \cdot \vec{d} &> 0
\end{aligned}
\tag{5.3}
$$

The first inequality makes the alternative logic to be located in the less critical area. The second inequality ensures that the alternative logic is located along the gradient direction of TW. Figure 5.15 shows an example of different AWs. $AW1$ is an available alternative logic, while we can skip $AW2$ and $AW3$ since they are likely to be futile or lead to a local minimum.

5.2.2.2 Timing Estimation

Timing calculation is another difficulty during the logic shift process since global STA is very time consuming and we cannot afford to call it during each process. To balance the accuracy and runtime, we adopt a local delay estimation scheme that can evaluate the change in circuit delay accurately in most cases with much less CPU time.

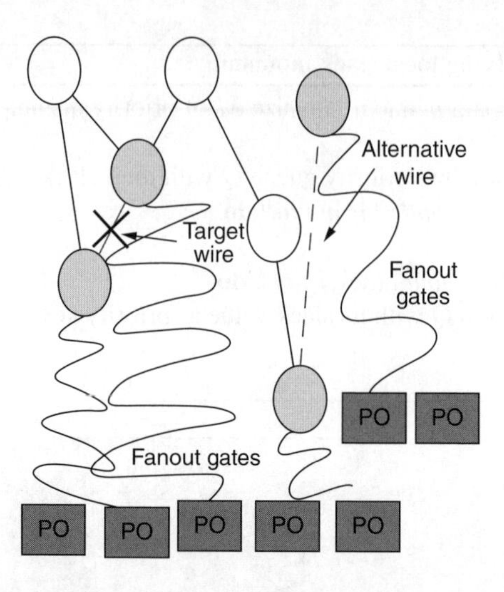

Figure 5.16 Fanout gates of TW and AW

After a rewiring process, we only update the arrival time of all fanout gates of the sources and the sinks of TW and AW. We define the local worst negative slack (LWNS) and local total negative slack (LTNS) of a set of gates as in the following to help evaluate the AW candidates.

Given a set of gates $G = \{g_1, g_2, \cdots, g_n\}$,

$$LWNS(G) = \min\{slack(p_t(g_i))\}$$

$$LTNS(G) = \sum_{i=1}^{n} slack(p_t(g_i)) \tag{5.4}$$

where $p_t(\cdot)$ denotes a gate set in which each gate is a primary output and belongs to the fanout cone of a gate g_i in G, such as the POs, as shown in Figure 5.16.

We will adopt an AW if

1. The LWNS of the rewired circuit is no worse than the original circuit;
2. The AW gains the most LTNS improvement among all AW candidates.

The first condition guarantees that the delay of the rewired circuit is not degenerated. The second condition helps in reducing the delay of all paths.

After an AW is accepted by the above procedure, a more accurate global STA is executed to verify whether the chosen AW is really improving the circuit delay. The circuit will be restored if the double check by the global STA fails.

The flow for removing a TW is shown in Algorithm 5.5.

5.2.2.3 Experimental Results

The proposed framework as well as a rewiring engine is implemented in C++. We compare our algorithm with a previous work. Our initial placements were generated by Cadence SoC

Algorithm 5.5: Shifting logic for timing optimization

```
1  Input: Target wire t, slack distribution graph G;
2  begin
3      Compute gradient g⃗_t of t from G;
4      Call logic rewiring engine and collect all AW candidates for t;
5      for each AW candidate a do
6          Compute distance d⃗ between a and t;
7          if slack_a < slack_t or g⃗_t · d⃗ < 0 then
8              continue;
9          Temporarily remove t and add a with overlapping-free incremental placement;
10         Compute initial LWNS W and LTNST;
11         Update the arrival times of the fanout gates;
12         Compute updated LWNS W' and LTNST';
13         if W' < W then
14             continue;
15         Update best LTNS gain g and best AW candidate α;
16         Restore netlist and arrival times of the fanout gates;
17     if α is found then
18         Remove t and add α;
19         Perform global STA to double check if α addition improves circuit delay;
20         if Double check failure then
21             Restore netlist;
22 end
```

Encounter 10.1 based on the original synthesized netlist from the IWLS 2005 benchmark suite. The Cadence 180 nm generic library is used, and the circuit timing is estimated by STA with the same library. For simplicity, we only consider the timing of combinational circuits. And the output/input pins of a sequential cell are taken as the primary inputs/outputs.

To evaluate the proposed mountain-mover idea, we first run a more (traditional) greedy-like scheme where the most critical (slack peak) areas are eliminated first by shifting the related wirings to their less critical neighboring areas, and the result is shown in last column of Table 5.6. The second flow uses our novel framework where the boundary of the slack mountain (zero slack areas) will be rewired first to leave room for following bottom-up rewirings toward peak top areas. The result is shown from the 8th to the 12th columns. The proposed boundary-first optimization flow can gain more delay reduction (on average 14.1%) compared to the critical area first flow (9.2%), which demonstrates that this intuitive logic shifting scheme works better for eliminating more critical slacks. Why? We suspect that the greedy peak-top-first scheme is more vulnerable to be trapped in local minimums (valleys instead of sea areas). As shown in the "slack mountain maps" of the initial and optimized circuit for circuit mem_ctrl, Figure 5.17(a) and (b), slack mountains are effectively shoveled or smoothed down.

Table 5.6 Effectiveness of our algorithm compared to the traditional peak-first algorithm and a recent work

			Recent work				Our algorithms (boundary first)				Peak[a] first
			Performance		Area increase		performance		Area Increase		
Circuit	#cell	#net	%delay improvement	Time (s)	%wire	%cell	%delay improvement	Time (s)	%wire	%cell	%delay improvement
sasc	563	568	14.1	41	2.35	3.13	44.78	1.6	−10.85	6.36	30.0
spi	3,227	3,277	10.9	949	4.53	0.73	15.63	23.8	−1.01	1.15	10.4
des_area	4,881	5,122	12.3	503	1.09	0.31	13.26	119.3	0.32	1.43	8.1
tv80	7,161	7,179	9.1	1,075	2.50	0.17	8.99	67.1	−0.30	1.14	3.4
s35932	7,273	7,599	27.5	476	2.14	0.19	16.21	19.7	−1.12	0.62	14.3
systemcaes	7,959	8,220	13.9	748	0.89	−0.07	8.43	39.6	−0.32	0.98	4.0
s38417	8,278	8,309	11.7	481	0.68	−0.21	6.75	38.1	0.15	1.43	6.2
mem_ctrl	11,440	11,560	9.2	678	0.05	−0.02	18.33	391.9	−0.05	1.79	9.7
ac97_ctrl	11,855	11,948	6.3	245	0.44	0.02	1.77	7.5	0.19	0.24	1.8
usb_funct	12,808	12,968	12.2	605	0.30	0.06	12.92	56.0	−0.43	1.28	7.6
DMA	19,118	19,809	14.5	845	0.16	0.08	28.95	205.4	0.14	0.72	19.8
aes_core	20,795	21,055	6.4	603	0.13	0.01	3.15	74.1	1.21	0.47	2.4
ethernet	46,771	46,891	3.7	142	0.08	0.06	4.70	16.0	0.00	0.02	2.2
Average			**11.7**	569	1.18	0.34	**14.14**	81.5	−0.93	1.36	9.2
Ratio				**7×**				1			

[a]Peak first: the result of traditional flow which optimizes the most critical paths first.

We compare our algorithm with a recent work whose results are shown in the fourth to seventh columns of Table 5.6. Note that though the logic netlists used by us and the recent work are the same, a straight comparison is not appropriate here, as the placement engines used are different. But, since both placement tools are reasonable and comparable, we assume the delay improvement statistics are still useful and noticeable.

On average, our algorithm gains 2.4% more delay reduction compared to the recent work (14.1% vs 11.7%) with a speedup of at least 7 times. It is observed that our algorithm and the recent work performed very differently on some designs (e.g., on circuits sasc, mem_ctrl, DMA), and s35932 and systemcaes are optimized better using the recent work. This relatively large difference implies that perhaps the logic rewiring technique is exploring some optimization domains less explored by other logic restructuring techniques, and vice versa. It might be beneficial to look into some coupling integrations using different approaches together. The "Area Increase" columns in Table 5.6 show the wirelength and cell increase in percentages, while the fifth and ninth columns in Table 5.6 show the CPU times.

5.3 ECO Timing Optimization

Engineering change order (ECO) refers to the small-scale requirement changes that are typically made late in the chip design cycle. They are changes to fix design defects and/or to meet modified constraints. Spare cells are extra cells that are inserted and preplaced into the unused

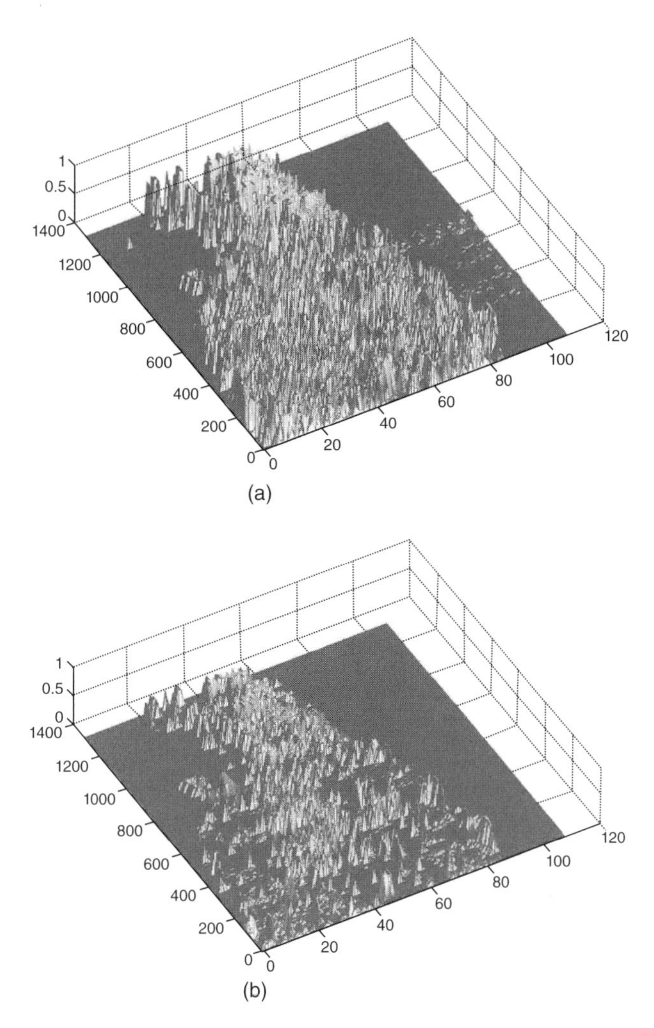

Figure 5.17 Slack mountain being smoothed down by timing optimization (valley areas are not shown for clarity). (a) Slack mountain of initial circuit; (b) slack mountain of optimized circuit

spaces of a chip. The type and number of spare cells vary from different chip designs and are usually determined by designers empirically. Spare cells are often used to perform ECO after the placement to fix/optimize a design incrementally. By changing metal connections only, spare cells can be used to substitute original committed cells (active cells), and the replaced active cells then can become new spare cells.

In this section, we propose two different algorithms to further optimize the timing of a circuit using spare cells. One algorithm does not touch any logic function of the circuit and thus can be attributed to the traditional physical optimization techniques, while the other one restructures the local circuit using rewiring and has the ability to further optimize the circuit even after the full optimization using traditional physical optimization techniques. The results show the ideal combination between these two physical and logical optimization techniques.

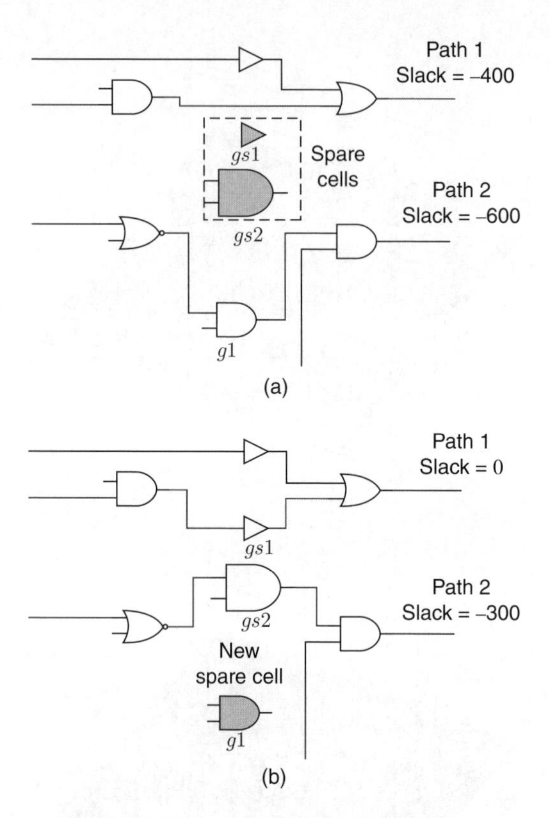

Figure 5.18 Example of ECO path optimization. (a) Before ECO optimization; (b) after ECO optimization

5.3.1 *Preliminaries*

Buffer insertion and gate sizing are two main techniques in ECO timing optimization, and Figure 5.18 shows an instance of timing optimization using these two operations. Figure 5.18(a) shows the original paths that violate the timing constraints. The buffer g_s1 and the AND gate g_s2 are spare cells. We can fix the timing violations of path 1 by inserting the buffer g_s1 to boost the driving speed. For path 2, the spare cell g_s2 of the same gate type but with larger driving capacity substitutes the active gate $g1$. After the changes made to the netlist, $g1$ becomes a new spare cell (which can be utilized again in later operations), while g_s1 and g_s2 are now active cells. The slacks on both paths are well improved.

Definition 5.1 *A timing path is defined, similarly, as a path between terminals or sequential elements, which can be primary input/output, latch, or flip-flop.*

Definition 5.2 *An ECO path is defined as a timing path that violates the timing constraints. The total negative slack (TNS) is a common metric to measure the gap between the actual arrival time and the required timing constraint.*

Definition 5.3 *An ECO path* NEGO-ROUT *operation is to rebuild the ECO path using either active cells on any ECO paths or spare cells with the same logic functions.*

Definition 5.4 *An ECO path restructuring operation involves the replacement of a/some active circuit cell(s) on an ECO path with appropriate spare cell(s) in the netlist while preserving the circuit's original functionality.*

Given the above definitions, the ECO timing optimization problem can be formally defined as follows: *Given a pre-placed netlist with spare cells inserted, the ECO timing optimization is to perform either physical operations or restructuring operations so that the resultant circuit satisfies the timing constraints.*

In this section, routing refers to the selection of suitable routing resources (gates or buffers) from a source node to a destination node. It is slightly *different* from the classic global or detailed routing in which physical geometries of the routing tree are decided and drawn. Each ECO path begins with the source node, which can be any primary input/sequential cell, and ends with the destination node, which can be any primary output/sequential cell. All circuit gates on the ECO paths plus the spare cells make up the available routing resources, and these resources can further be classified into different groups according to their logic functions: cells having the same logic function are gathered in one group and can be used interchangeably on ECO paths.

5.3.2 NEGO-ROUT *Operation*

In this section, we present a negotiation-based routing algorithm to perform the so-called NEGO-ROUT operation. The cost function used in the algorithm is similar to that in McMurchie and Ebeling (1995). It is able to negotiate the resource competition among multiple ECO paths. We summarize the algorithm in Algorithm 5.6.

Different from other previous works, our algorithm allows a gate initially shared by multiple ECO paths and the process will continue until there is no congestion in any path. In the beginning, our algorithm ignores any congestion and reroutes the paths in order to obtain the best slack even though the selected gates are shared by many ECO paths. Afterward, selecting a shared gate will be discouraged by a congestion penalty φ_i. The penalty will be increased as the algorithm iterates. The penalty will guide the algorithm to select gates with a little larger delay but shared by fewer paths.

We start rerouting an ECO path from the destination node and select the new gates for the path one after another until the source is reached. We record all possible solutions from the destination node to the current reaching node of the path. We also collect all available candidates for the next node of the path. We then combine each solution and candidate gate to generate new solutions. When we are done with the whole ECO path, we will choose the solution with the best cost as the final NEGO-ROUT solution of the ECO path.

The cost of a new solution NS obtained by combining solution P and next candidate gate g_i is calculated as below:

$$cost(NS) = cost(S) + d_i + \varphi_i \tag{5.5}$$

where $cost(S)$ is the cost value of the solution S, d_i is the delay of the candidate gate g_i, and φ_i is the gate congestion penalty used to prevent selection of shared gates.

Algorithm 5.6: ECO path NEGO-ROUT operation

```
1  Input: Initial ECO paths;
2  begin
3      break up all ECO paths and free up all routing resources;
4      route all ECO paths ignoring legalization;
5      while shared resources exist do
6          for each ECO path i (re-route one path) do
7              rip up ECO path i;
8              initialize solution set SS;
9              for each gate j on path i do
10                 collect all available candidates for gate j;
11                 initialize temp solution set TSS;
12                 for each candidate k do
13                     for each solution s in SS do
14                         create new solution ns by combining s and k;
15                         push ns into the TSS at cost cost(s) + d_k + φ_k;
16                 copy TSS to SS;
17                 prune solutions in SS;
18             reroute path i using the solution which has best cost;
19             update shared count and accumulated penalty φ_i for path i;
20 end
```

We follow the gate and wire delay estimation discussed in Ho et al. (2010). We look up the table of the Synopsys library where the value of driving capacitance and input transition time of the next gate are needed. With the *loading dominance* property, we can estimate a delay well by using the input transition time obtained in last iteration and recompute the wire loading as follows:

$$C_W(g_i) = \left(\sum_{g_j \in FO(g_i), g_j \notin path} (|x_i - x_j| + |y_i - y_j|) + |x_p - x_i| + |y_p - y_i| \right) \times \phi \tag{5.6}$$

where g_i is the next candidate gate and g_p is also on the path and is driven by g_i. $FO(g_i)$ includes all fanout gates of g_i. As g_p may differ from the original gate on the path, we must use the updated location of g_p to compute the wire loading.

The congestion penalty φ_i is computed as follows:

$$\varphi_i = (cost(S) + d_i + h_i) \times p_i \times \alpha_n \tag{5.7}$$

where h_i is the accumulated history penalty, which is the sum of the congestion penalties in previous iterations. The factor h_i can help the algorithm to jump out of infinite loops when there are multiple paths competing for one available gate with the same sharing count. The theory of h_i can be found in McMurchie and Ebeling (1995). p_i is the share count for the gate,

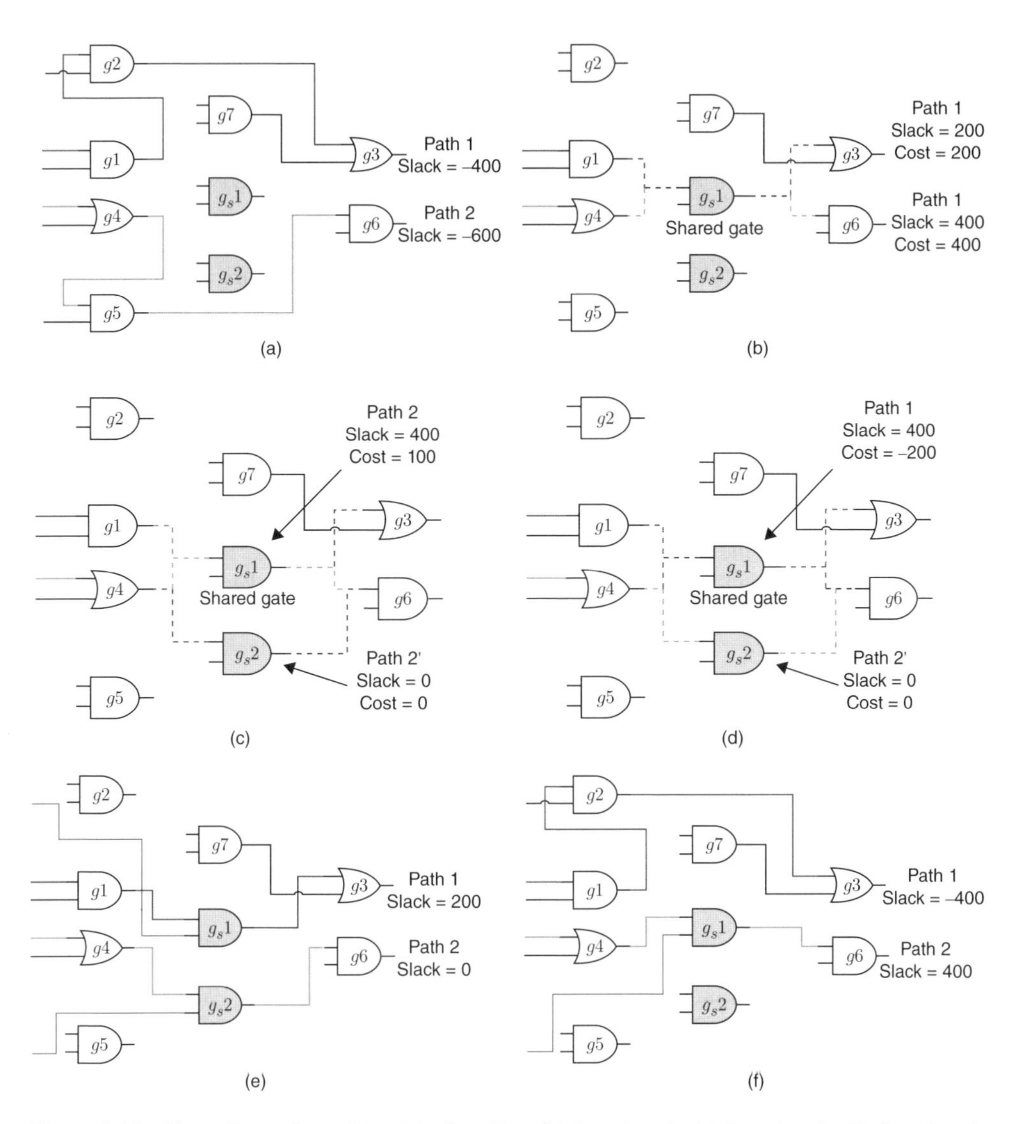

Figure 5.19 NEGO-ROUT flow. (a) original netlist; (b) iteration 1; (c) iteration 2; (d) iteration 3; (e) NEGO-ROUT result; (f) DCP result

and α_n is the factor of penalty, which increases with the iteration number n. If the gate is not shared currently, p_i will be zero and thus the penalty is also zero.

Figure 5.19 shows an example of solving the problem using the congestion penalty and illustrates the ability of global optimization among multiple paths competing for gate resources. We just use slack taking the role of delay to make it more appealing visually. Figure 5.19(a) shows the original netlist that contains two ECO paths. For path 1, substituting $g2$ by g_s1 can obtain positive slack, while both g_s1 and g_s2 can be used to take the place of $g5$ for path 2 to get positive slack. The NEGO-ROUT algorithm will just select the paths with the best slack for the first iteration shown in Figure 5.19(b). The congestion penalty begins working after the first iteration, and the solution cost of path 2 by selecting g_s1 becomes 100 and -200 for iterations 2 and 3 shown in Figure 5.19(c) and (d). The cost is always 0 when selecting g_s2 since g_s2 is not shared all the time and the cost is just equal to the slack. In the third iteration, the cost for selecting g_s2 is better than that for g_s1, and thus g_s2 is chosen to be the new active gate on path 2.

Figure 5.19(e) shows the final netlist after NEGO-ROUT operation, and both paths have positive slacks. Figure 5.19(f) shows the result produced by the DCP algorithm, which optimizes ECO paths one after the other. In the example, the DCP algorithm will first optimize path 2 because of its larger negative slack and use g_s1 to get the best solution, which prevents the further optimization to path 1.

5.3.3 Path-Restructuring Operation

In this section, we will explain how we apply logic rewiring to restructure the ECO paths. We always restructure the unlocked path with the largest negative slack. If the paths meet the timing constraint or cannot be optimized already, the paths will be locked. The restructuring operation will stop when all ECO paths are locked.

When we restructure a chosen path, we begin with an STA and calculate the slacks to all nodes in the circuit. For each wire on ECO path as the target removal wire, we use the logic rewiring technique to identify the alternative wire.

When we remove the wire on the ECO path and merge the alternative wire into the netlist, we can just compare the new slack values of the target and alternative destination gates with the original ones. As the slack of an ECO path is always equal to the slack of the node among the paths, TNS will be improved if the slack becomes better for the sum of the target and alternative gates. If the TNS is not improved, we cancel the rewiring operation and restore the netlist.

Figure 5.20 shows an example of restructuring the ECO path by rewiring. Figure 5.20(a) is the original ECO path, and by rewiring we find that the gate $g8$ can be replaced by the spare gate g_s2. After restructuring, the circuit is shown in Figure 5.20(b). We can see, obviously, from the figure that the ECO path has fewer gates and thus better timing. The greatest benefit of the restructuring operation is that the original gate may be replaced by a spare gate that has a different logic type (note that $g8$ is an OR gate while g_s2 is an AND gate), which is not allowed in any physical optimization technique.

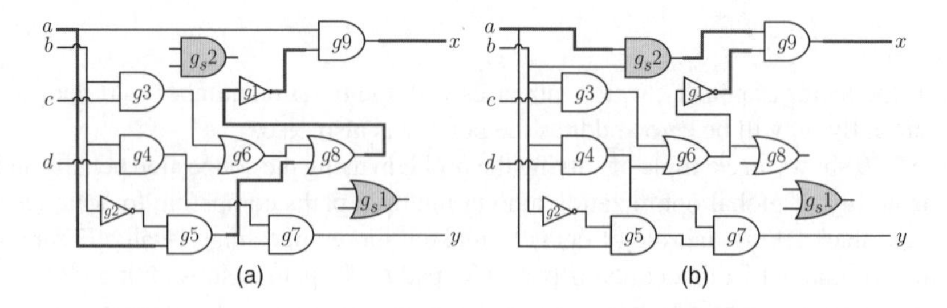

Figure 5.20 RESTRUCTURING flow. (a) Original netlist; (b) rewired netlist

We summarize the detailed flow to restructure a chosen ECO path in Algorithm 5.7.

Algorithm 5.7: ECO path restructuring operation

```
1  Input: Initial ECO paths;
2  begin
3      select the ECO path p = ⟨g_0, g_1, ..., g_n⟩;
4      for each wire w_t = (g_i, g_{i+1}) on p do
5          remove w_t;
6          compute the variation in the slack of g_i;
7          Δslack_tar = slack_new(g_i) − slack_old(g_i));
8          if Δslack_tar < 0 then
9              break;
10         use logic rewiring to find the alternative wires for w_t;
11         for each alternative wire w_a do
12             collect candidates from spare cells as an alternative gate;
13             for each candidate k do
14                 use candidate k as the alternative gate;
15                 compute the variation in the slack of g_k;
16                 Δslack_alt = slack_new(g_k) − slack_old(g_k));
17                 if Δslack_alt + Δslack_tar > 0 then
18                     update timing information by STA;
19                     if TNS_new < TNS_old then
20                         return the changed netlist;
21                 restore the alternative netlist;
22         restore the target netlist;
23 end
```

The combination of the two different algorithms is not a difficult work. Figure 5.21 shows the whole flow of the framework. The framework takes a circuit netlist with annotated timing requirements. ECO paths are then identified using the STA tool with a standard cell library in the Synopsys Liberty Library format. All ECO paths are optimized by NEGO-ROUT and logic restructuring operations successively.

In the ECO path NEGO-ROUT phase, we iteratively break up and reroute all ECO paths. A gate is no longer confined to be used for a specific path; instead, it is open up as an available routing resource for all paths.

In the ECO path restructuring phase, we repeatedly pick the ECO path with the large negative slack and perform logic rewiring to break the path. The recently published rewiring scheme Yang et al. (2010) is used to find alternative wires for target wires chosen on the ECO paths. This maximizes the number of candidates during the timing operations. Since the gates on different paths are changed, we have to update the timing estimation by STA after every rewiring operation. Under the assumption of loading dominance and shielding properties, the update can be completed in a much shorter time. If the TNS of the netlist cannot be improved, we will lock the path, denoting that this path cannot be further optimized, and restore the original

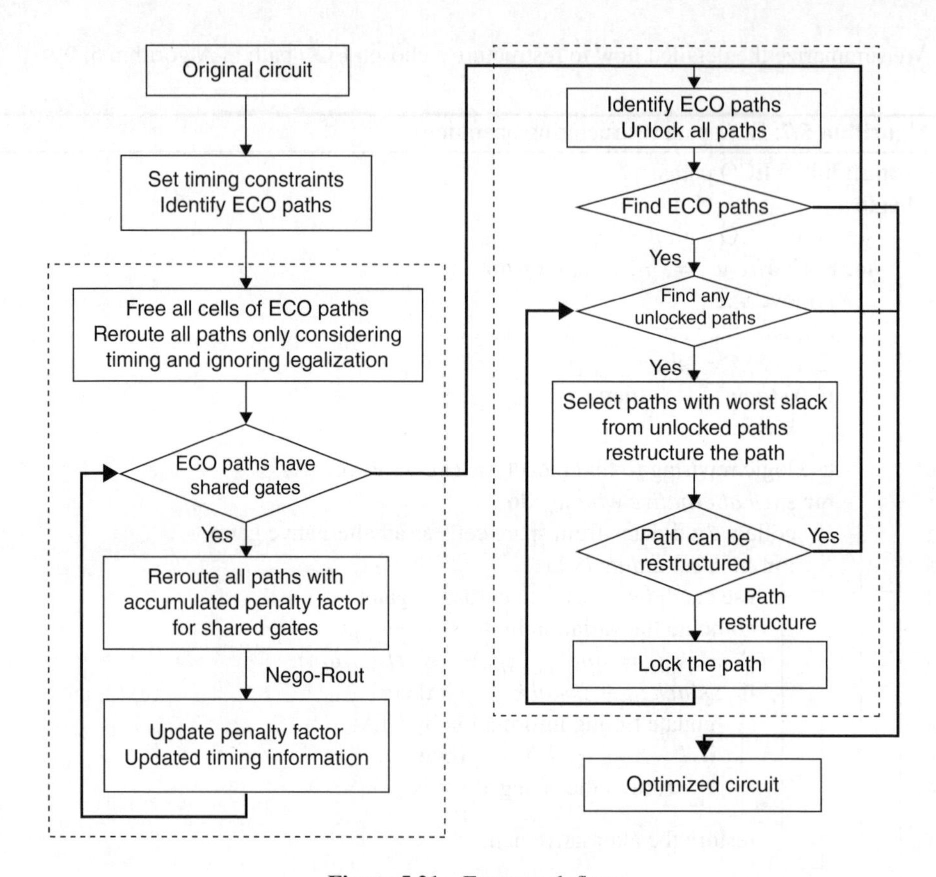

Figure 5.21 Framework flow

netlist. Then we continue to choose another path from unlocked paths to be optimized by the logic rewiring operation. If the TNS is improved, we shall update the timing information using STA and re-identify the ECO paths. The process continues until all ECO paths meet the timing constraints or all paths have been locked.

5.3.4 Experimental Results

We use the MCNC and ITC benchmarks, which include the logic information needed by the logic restructuring operation to test the power of our algorithms. We also run the previous DCP algorithm under the same benchmarks as a comparator. The benchmarks were mapped to a standard cell library in Synopsys Liberty format. Placement was then generated by the academic placer CAPO Roy et al. (2006). Nearly 10% of the gates in the benchmark are spare cells, which are distributed evenly in the placement.

In this ECO problem context, both buffer insertion and gate sizing operations can be treated similarly and integrated in the same algorithm flow (as in Chen et al. (2007) and our framework). Without loss of much generality, gate sizing operations will be represented to test the power of using DCP and our methods for comparison. To test the capacity limits of both schemes, each path is assigned a negative slack that is 10% of each original ECO path delay. Then we applied our algorithm and DCP schemes to fix the timing violations. Table 5.7 shows the experimental result. Columns 2–5 describe the basic statistics of the benchmarks used: total number of gates (excluding spare cells), total number of spare cells added, and total number of ECO paths, of the circuit netlist. Columns 6–11 show the optimized TNS and number of ECO paths by our method. Columns 6 and 7 show the results by using NEGO-ROUT operation, while columns 8–11 show the results by using both NEGO-ROUT and restructuring: columns 6 and 9 are the TNS after optimization; column 10 is the value of TNS reduction by the optimization; columns 7 and 11 are the runtime when we perform NEGO-ROUT or both NEGO-ROUT and restructuring; column 8 shows the number of ECO paths remaining. The last four columns show the optimization results by the DCP algorithm.

The results suggest that our operations can reduce 18% more TNS compared to DCP, and the final TNS of our algorithm is only 50% of the DCP result. The runtime of our algorithm is a little longer than that of DCP (ours takes 1.33× runtime than DCP). However, given the fact that ECO is a very last EDA stage, a longer run to search for a workable solution is definitely still desirable, as it still save much time compared to starting over again or solving it manually.

The results shown in column 6 suggest that our NEGO-ROUT can reduce no less TNS except for C3540. In NEGO-ROUT, a timing-optimal but illegal netlist will be found first. Then we resolve the shared gates considering the competition among multiple ECO paths gradually through iterations. In DCP, the author simply tries to optimize each ECO paths one by one sequentially and almost independently. So the two schemes have two totally different thoughts, and the results suggest that NEGO-ROUT will gain more TNS reduction in most of cases.

The results shown in column 9 suggest that restructuring can further reduce 10% of the TNS on average. Restructuring can change the topology of the netlist and break the ECO paths, which may create more room for the solution space. The restructuring operation is nearly independent of the results produced by either NEGO-ROUT or DCP.

5.4 Area Reduction in FPGA Technology Mapping

Logic perturbations inside a look-up-table (LUT) can be considered to be completely free since a K-LUT is able to implement any K-variable function. Logic changes inside an LUT impose no penalty in both logic usage and delay on FPGA architectures. However, in most technology mapping algorithms, the logic information of the circuit is ignored and the circuit is modeled as a directed acyclic graph (DAG) during the mapping. Most approaches improve area and delay by applying different heuristics and algorithms (bin-packing, binate covering) on depth-optimal mapping solutions without any consideration to the logic transformation of the subject circuit.

In this section, a generic approach, which allows the combination of logic rewiring and technology mapping technology, to further improve depth and area of the mapping solutions will be introduced. In experiments, various best known technology mapping algorithms were tested, and the results are promising: substantial depth and area improvement can be further achieved for different technology mapping algorithms. Studies from Cong and Minkovich (2007) show that technology-independent logic synthesis can perform much worse (70 times larger) than the optimal solution. Our work thus aims to provide a general method to tune logic synthesis well for technology mapping at the expense of a small runtime cost. We use rewiring algorithms as our basic logic perturbation tool in the suggested method.

Table 5.7 Comparison of our algorithm and the DCP algorithm

Benchmark information					NEGO-ROUT		NEGO-ROUT+RE-STRUCTURING				DCP			
Circuit name	#gates	#spare cells	#ECO paths	TNS (ns)	TNS (ns)	Runtime (s)	#ECO paths	TNS (ns)	Gain	Runtime (s)	#ECO paths	TNS (ns)	Gain	Runtime (s)
alu4	775	75	7	−9.45	−1.64	4.1	2	−0.26	9.19	13.2	5	−3.64	5.81	1.6
apex2	15,983	1,452	3	−6.42	0	11.0	0	0	6.42	11.5	0	0	6.42	11.4
apex6	809	75	72	−20.23	−3.00	10.5	18	−1.95	18.29	14.4	39	−4.87	15.36	3.2
b14	5,992	576	31	−69.67	−12.25	130.3	12	−11.27	58.40	157.8	13	−18.76	50.89	36.3
b15	9,890	900	48	−77.49	−13.19	80.3	22	−13.19	64.30	84.1	25	−14.85	62.64	36.1
b20	13,066	1,224	128	−256.73	−40.38	430.0	68	−29.10	227.60	970.4	87	−66.34	190.30	800.7
b21	13,163	1,224	104	−283.71	−34.32	852.2	38	−31.59	252.11	1,077.0	47	−58.12	225.59	449.6
b22	19,411	1,764	126	−304.10	−56.90	432.7	51	−46.40	257.70	1,292.9	80	−96.54	207.50	1,639.8
C1355	513	48	8	−9.35	−1.23	13.2	4	−1.23	8.12	15.1	7	−3.10	6.25	2.3
C1908	459	48	15	−10.79	−1.48	6.7	8	−1.35	9.44	11.9	13	−4.74	6.06	3.0
C3540	1,171	108	11	−14.11	−3.62	12.5	2	−0.81	13.30	14.0	5	−3.59	10.52	2.5
C6288	2,557	243	19	−32.65	−3.82	275.9	7	−3.56	29.09	285.2	14	−5.66	26.99	25.1
DES	3,721	363	94	−92.73	−12.06	93.3	24	−11.07	81.66	143.5	45	−28.27	64.46	73.6
misex3	520	48	10	−7.63	−1.65	4.3	4	−1.65	5.98	14.7	6	−1.87	5.76	1.7
x3	764	75	75	−17.86	−6.03	5.3	44	−6.03	11.83	10.2	55	−6.58	11.28	2.8
Total	88,794	8,223	751	−1,212.88	−191.56	2,362.3	304	−159.45	1,053.43	4,114.5	441	−316.92	896.85	3,088.9
Ratio					0.60	0.76	0.69	0.50	1.18	1.33	1	1	1	1

TNS: total negative slack; *gain*: reduction of circuit TNS.

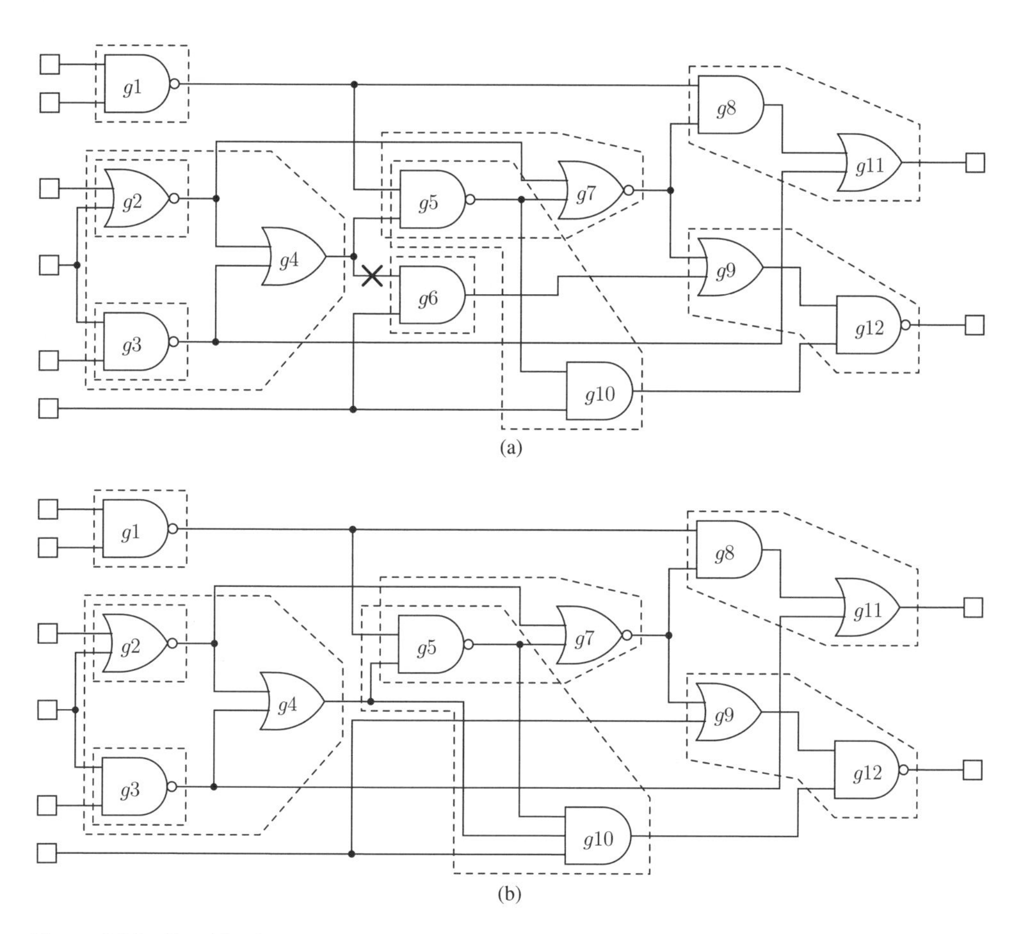

Figure 5.22 Rewiring improvement upon already optimal graph-based tech map. (a) Optimal mapping without logic perturbation: depth = 3, area = 9; (b) improved mapping after logic perturbation: depth = 3, area = 8

Two optimization routines are proposed in this paper. The first routine optimizes the logic level of the circuit, while the second routine optimizes the resultant area required for mapping the input circuit.

Our depth optimization focuses on removing wires on the critical path(s) of the circuits. The idea is based on the observation that if the number of levels of logic gates in the subject circuit is reduced, so may the depth of the mapping. We used rewiring to optimize the subject circuit so as to reduce its levels of logic gates. By performing rewiring transformations on the gates along the critical path and restructuring the path, we can bring down the number of logic levels.

To illustrate how rewiring can improve the mapping area beyond the combinatorial limit, consider the example with $K = 3$ in Figure 5.22. The original mapping by DAOMap, denoted by dashed boxes, is already optimal in depth and area if the structure of the circuit cannot be changed. Now suppose we replace $g4 \rightarrow g6$ with $g4 \rightarrow g10$; gate $g6$ is then removed and $g10$ is expanded to be a three-input AND gate, which can be implemented by an LUT with $g5$ together. Area is reduced by 1, while the depth is preserved.

The area optimization in this approach is called *incremental logic resynthesis* (ILR). ILR provides a greedy search on the logic perturbations, which can bring incremental area reductions. The target wire candidates are first ranked based on topological order or on the ranking procedure we propose in this paper. We then perturb the subject circuit with a target wire removal and alternative wire addition mechanism and then evaluate this perturbation based on the resultant depth and area. The perturbation will be accepted when a smaller area is achieved. A series of perturbations are applied in a ranked manner to allow efficient reductions of the mapping area. A number of heuristics will be suggested to search for useful target wires to be removed and alternative wires to be added.

ILR is generic and allows any technology mapping algorithms to be coupled with ease. While our logic level optimization brings down the circuit delay with very little area penalty, the incremental logic resynthesis technique allows deeper logic perturbation search and compensates the area penalty during delay optimization. This, once again, proves the usefulness of logic perturbations, which work beyond the combinatorial limit set by conventional technology mapping algorithms.

In the following, for simplicity, we assume all LUTs are of the same size, and the size of an LUT is denoted by K. A circuit is K-bounded if for every gate/LUT v, $|input(v)| \leq K$. The *unit delay* model will be used for delay estimation. The depth of a circuit is the largest level of LUTs among all the paths from primary inputs or outputs of flip-flops to primary outputs or inputs of flip-flops.

5.4.1 Incremental Logic Resynthesis (ILR): Depth-Oriented Mode

We are interested to solve the depth optimization problem for LUT-based FPGA technology mapping: that is, to minimize the number of level of LUTs to cover the l-bounded ($l \leq K$) subject circuit.

The depth of the original unmapped 2-bounded subject circuit influences the depth of the mapped solution directly, as the technology mapping process can be seen as packing gates in the subject circuit into K-LUTs. We shall describe a method of utilizing logic perturbation based on rewiring to modify the subject circuit aiming at reducing its depth. The optimized subject circuit is then used as input to the technology mapping algorithm. And, hopefully, a mapping solution with a better depth will be achieved. Hereafter, we will abbreviate ILR in the depth-oriented mode as *depth-ILR* for simplicity.

The critical gates in a circuit can be easily found by performing a topological search. Since removing an input wire to a two-input gate also removes the gate itself, by targeting the input wires of critical gates we can remove them by adding alternative wires elsewhere. Removing a critical gate can potentially reduce the depth since the number of gates along the longest paths can be reduced by 1 if the gate is a part of all longest paths or there is only a single longest path.

Depth-ILR is basically perturbation-based. For each perturbation pass, we first compute the current set of critical gates. For each input wire in the critical gates, its alternative wires are found by the rewiring algorithm. We then greedily choose an alternative wire pair from all pairs that can achieve a positive gain; that is, the depth can be reduced by 1 after the rewiring operation. If no alternative wire pair can achieve a positive gain, one is chosen randomly from zero-gain wire pairs. The chosen alternative wire pair is then used to transform the circuit by adding the chosen alternative wire and removing the target wire. To prevent the algorithm from oscillating between a few local minimum solutions, two heuristics are used: (i) if the number of consecutive zero-gain perturbations exceeds the user input parameter n, an alternative wire pair that can produce the largest increase in depth is chosen so as to perturb the Boolean network to escape from a local minimum; (ii) the set of previously added wires is well recorded so that if a zero-gain wire pair has to be chosen because of the absence of any positive-gain choice, wire pairs having the target wire added before will be assigned a lower priority than those that are not in the set. The total number of perturbation passes is limited by another input parameter n. The overall algorithm is shown in Algorithm 5.8. The optimized subject circuit is then used as the input to the technology mapping algorithm.

Algorithm 5.8: Outline of depth-ILR

```
1  begin
2      for perturbation i = 1 → n do
3          for each input w_t of critical gates do
4              A ← REWIRING(w_t);
5              if ∃t_p = (w_t, w_a ∈ A) s.t. gain(t_p) > 0 then
6                  G ← TRANSFORM(G, t_p);
7              else
8                  randomly choose t_p where gain(t_p) = 0;
9                  G ← TRANSFORM(G, t_p);
10 end
```

5.4.2 Incremental Logic Resynthesis (ILR): Area-Oriented Mode

We are interested to solve the depth-preserving area minimization problem for LUT-based FPGA technology mapping: that is, to minimize the number of K-LUTs used to cover the l-bounded ($l \leq K$) subject circuit while keeping the depth of the mapped circuit optimal. Any gates that lie on the critical paths are called *critical gates*.

Most of the current technology mapping algorithms do not consider local logic transformations for either depth or area optimization. Optimality studies (Cong and Minkovich 2007, Ling et al. 2005) show that by utilizing the logic information, one can improve beyond the optimization limit imposed by DAG-based technology mapping algorithms. This motivates us to incrementally resynthesize the subject circuit using logic transformations so that a smaller mapping area can be achieved in the mapped solution under the depth-optimality constraint. Hereafter, we will abbreviate ILR in the area-oriented mode as *area-ILR* for simplicity.

The basic idea in area-ILR is to incrementally restructure the subject circuit so that a smaller mapped area is used. Before the area optimization process starts, a ranking of the target wires will be obtained first.

The ranking procedure is important for reducing the runtime of the whole algorithm. The details of the ranking procedure (`ILR_WIRE_RANK`) will be explained in Section 5.4.2. This procedure produces a ranked list of wires for perturbations. We extract the candidate target wire one by one and find the alternative wire set for each target wire. We then apply each transformation to the network and try to map it using the chosen technology mapping algorithm. Supposing the transformed subject circuit is mapped with less number of LUTs and the

Algorithm 5.9: Outline of area-ILR

```
1  begin
2      L_{w_t} ← ILR_WIRE_RANK(E(G_S));
3      foreach wire w_t ∈ L_{w_t} do
4          A ← REWIRING(w_t);
5          foreach w_a ∈ A do
6              G'_S ← TRANSFORM(G_S, (w_t, w_a));
7              G'_M ← TECH_MAP(G'_S, K);
8              if d(G'_M) ≤ d(G_M) and |V(G'_M)| < |V(G_M)| then
9                  G_S ← G'_S;
10                 G_M ← G'_M;
11                 break;
12 end
```

depth of the mapped circuit is reduced or unchanged, we would accept the transformation and update both the subject circuit and the mapped circuit.

The algorithm examines the set of available transformations and greedily takes the transformation when it can reduce the number of LUTs. The process reduces area incrementally, giving a series of transformations that bring area reductions, until no more reduction can be found. The final mapped circuit will be returned.

As area-ILR is an incremental optimization algorithm, we can terminate the optimization process once the area objective or constraint is satisfied. Our technique will greedily reduce the area of the design until it can be implemented on the device, and it will preserve the depth of the design to avoid deterioration of performance. Thus it can be the last resort to allow more area-hungry designs to be implemented on resource-limited devices.

The order in which the target wire is tested significantly affects the efficiency of the whole area-ILR. It is desirable to have more area-reducing transformations at the beginning of the list so that we have a higher chance to find it during the sequential evaluation during the ILR algorithm. We suggest two heuristics for ranking the wires.

First of all, we observed that the wire closer to the primary input tends to have higher possibilities to reduce the area of the mapped circuit when it is chosen to be the target wire. This is because the perturbations closer to PIs would affect a wider part of the mapped circuit, especially when the gate duplications and sharings are more frequent there. Based on the statistics we collected, very few useful transformations can be found at one or two levels from the primary outputs; on the contrary, around 70% of useful transformations can be found within 5–6 levels from primary inputs.

We give higher ranks to the wires closer to PIs. In our algorithm, we first sort the gates in topological order and form an initial list of target wires by taking fanouts of the sorted gates. As the optimization proceeds, the target wires on the initial ranked list are evaluated one by one, and the algorithm stops the evaluation at the point where it finds a useful transformation to reduce the number of LUTs in the mapped circuit. We observed that the wires evaluated not useful in area reduction during this iteration will likely remain not useful during next iterations. Thus we give higher ranks to untested target wires and lower ranks to tested but unused target wires. We record the position on the list where we stop in this iteration. In the next iteration, we continue at the same or closest position. In our experiments, the ranked wire list is implemented as a circular linked list, which makes position recording easier. On average, we can reduce 60% of runtime by having this ranking procedure at the beginning of each iteration. This is the main ranking procedure in our experiments.

5.4.3 Experimental Results

MCNC benchmarks were used in many of the following experiments. They were first processed with *script.algebraic* and decomposed into simple two-input gates (AND, OR, NOT) using *mcnc11.genlib*. Unless specified, in the experiments for area minimization, the mapped depth of the optimized circuits are the same as the initial ones.

We first compared the LUT depth reduction with FlowSYN, which applies logic reconstruction based on binary decision diagram (BDD) decomposition to reduce mapped depths. We processed 20 MCNC benchmarks with the logic level reduction routine in our framework. FlowMap is used to finalize the mapping. The same set of benchmarks is processed with FlowSYN. Because of space limitation, we show only those benchmarks with depth reduced by either FlowSYN or our approach. The results are shown in Table 5.8.

Experimental results show that our logic level reduction method is more stable in terms of reducing the mapped depth in more circuits; our method can reduce the mapped depths in 9 out of 11 benchmarks shown, compared to only 3 improved by FlowSYN. We achieve 11.3% LUT depth reduction compared to 6.0% by FlowSYN. However, the area used in this approach may rise slightly due to logic duplications. Therefore we also apply ILR in the area-oriented mode as a remedy. The last three columns of Table 5.8 show the final result after depth-ILR. Except the circuits *9sym-hdl* and *5xp1*, all final mapped areas are smaller than that in FlowSYN. The area reduction is even more significant with the large benchmarks such as *alu4* and *C3540*. The results show that, besides a significant depth reduction of 11.3% (compared to FlowMap), the area can also be reduced by 6.1% (compared to FlowSYN).

DAOMap provides area-minimized and depth-optimal mapping solutions and is currently one of the best technology mapping algorithms. We would like to see if we can further reduce area produced by DAOMap when logic perturbation is allowed.

To illustrate the performance of area-ILR on different values of K, we conducted the experiments with $K = 4$ and $K = 6$, which are the most common values in commercial FPGAs nowadays. Table 5.9 shows the area reduction with DAOMap with $K = 4$. We compared the number of LUTs reduced by the initial mapping results with DAOMap. The percentages of LUT reduction over initial mapping results are shown in column 6. The experimental runtime in minutes and the runtime per LUT reduction are shown in columns 7 and 8, respectively.

Among all benchmarks, the depths of the mapped circuits are the same. Despite the fact the mapping area is already largely reduced by efficient heuristics in DAOMap, when logic perturbation is used substantial area reduction is still possible. Area-ILR can reduce 11.4% of LUTs upon DAOMap's mapped solutions, and in cases like *misex3* and *C6288*, over 20% of mapping area is reduced.

The performance of area-ILR with $K = 6$ is presented in Table 5.10. On several benchmarks such as *alu4*, *C5315*, and *C7552*, area-ILR also reduces the depths of the mapped circuit by 1. The area-reducing power is close to that in $K = 4$: 12.2% of mapping area is reduced on average. In practice, the proposed approach is versatile and can be extended to any practical values of K.

Table 5.8 Results of depth-ILR followed by area-ILR with FlowSYN ($K = 5$)

			FlowMap		FlowSYN		
Circuit	#PI	#PO	Dep.	#	Dep.	#	$D.r.\%$
9sym-hdl	9	1	5	20	4	15	20
5xp1	7	10	3	30	2	21	33.3
C1908	33	24	8	138	7	133	12.5
C880	60	26	8	147	8	146	0
alu2	10	6	9	174	9	164	0
C2670	233	140	8	195	8	195	0
x3	135	99	5	252	5	253	0
apex6	135	99	6	271	6	272	0
alu4	14	8	11	313	11	299	0
C3540	50	22	11	472	11	463	0
C6288	32	32	24	1,348	24	1,167	0
Total			98	3,360	95	3,128[a]	

	LR + FlowMap			LR + FlowMap + AR			
Circuit	Dep.	#	$D.r.\%$	Dep.	#	$D.r.\%$	$A.r.\%$
9sym-hdl	4	17	20	4	17	20	−13.3
5xp1	3	29	0	3	27	0	−28.6
C1908	8	138	0	8	132	0	0.8
C880	7	145	12.5	7	137	12.5	6.2
alu2	8	185	11.1	8	160	11.1	2.4
C2670	7	189	12.5	7	187	12.5	4.1
x3	4	255	20	4	235	20	7.1
apex6	5	267	16.7	5	249	16.7	8.5
alu4	9	301	18.2	9	264	18.2	11.7
C3540	10	468	9.1	10	429	9.1	7.3
C6288	23	1,163	4.2	23	1,126	4.2	3.5
Total	88	3,157		88	2,963		6.1

Only circuits with depth reduced in either FlowSYN or our method are shown in this table.
[a]Produced by RASP executable from UCLA.
LR: Reduction in depth-ILR. **AR**: Reduction in area-ILR.
$D.r.\%$: Depth reduction percentage (compared with FlowMap).
$A.r.\%$: Area reduction percentage (compared with FlowSYN).

IMap (Manohararajah et al. 2006) is an iterative technology mapping algorithm that adjusts its cost functions between iterations and gradually reduces the area or depth of the mapping. The nature of IMap is quite different from that of DAOMap, and in some cases IMap can produce better results than DAOMap in even shorter execution time.

We also performed another experiment to test the optimization power of area-ILR on improving mapping area with IMap. The area flow score heuristic is used in this experiment.

Table 5.9 Result of area-ILR over DAOMap ($K = 4$)

	DM		DM+ILR			
Circuit	Dep.	#	Dep.	#	% R.	mins
alu2	12	158	12	130	17.7	12.1
alu4	15	280	15	239	14.6	47.2
apex6	7	240	7	220	8.3	8.4
duke2	5	153	5	135	11.8	5.4
misex3	9	223	9	175	21.5	19.3
rot	10	237	10	221	6.8	5.8
term1	5	70	5	59	15.7	1.6
x3	6	243	6	224	7.8	6.1
C1355	6	80	6	78	2.5	3.7
C1908	9	133	9	122	8.3	2.7
C2670	10	182	10	181	0.5	0.3
C3540	14	352	14	323	8.2	85.5
C5315	9	517	9	507	1.9	8.2
C6288[a]	33	979	32	649	33.7	384.8
C7552	12	483	12	427	11.6	240.5
Total		4,330		3,690	14.8	
Average					11.4	

Ranking: Topological order; [a]Depth reduced.
DM: # of LUTs after DAOMap.
+ILR: # of LUTs after ILR area-oriented mode.
% Red: Reduction percentage. **mins**: Runtime in minutes.

Table 5.10 Result of area-ILR over DAOMap ($K = 6$)

	DM		DM+ILR			
Circuit	Dep.	#	Dep.	#	% Red.	mins
alu2	8	110	8	91	17.3	21.4
alu4	9	192	9	164	14.6	98.2
apex6[a]	5	181	4	161	11.0	9.3
duke2	4	111	4	101	9.0	6.6
misex3	6	138	6	109	21.0	22.8
rot	6	189	6	172	9.0	9.3
term1	4	46	4	41	10.9	2.4
x3	4	171	4	165	3.5	7.0
C1355	4	64	4	62	3.1	27.7
C1908	6	85	6	80	5.9	11.0
C2670	6	115	6	108	6.1	81.8
C3540	9	277	9	246	11.2	236.4
C5315	6	296	6	284	4.1	138.3
C6288	21	584	21	471	19.3	758.3
C7552	8	340	8	303	10.9	999.7
Total		2,899		2,558	11.8	
Average					10.5	

Ranking: Topological order. [a]Depth reduced.
DM: # of LUTs after DAOMap.
+ILR: # of LUTs after ILR.
% Red: Reduction percentage. **mins**: Runtime in minutes.

Table 5.11 Result of Area-ILR over IMap ($K = 4$)

	IM		IM+ILR			
Circuit	Dep.	#	Dep.	#	% Red.	mins
alu2	12	150	12	133	11.3	6.0
alu4[a]	15	272	14	255	6.3	21.3
apex6[a]	7	237	6	219	7.6	7.4
duke2	5	150	5	141	6.0	3.4
misex3	9	210	9	176	16.2	8.6
rot[a]	10	237	8	234	1.3	5.7
term1	5	71	5	56	21.1	1.1
x3	6	241	6	219	9.1	5.4
C1355	6	78	6	78	0.0	3.6
C1908	9	128	9	121	5.5	2.2
C2670	10	183	10	175	4.4	21.4
C3540	14	356	13	346	2.8	39.3
C5315	9	519	9	504	2.9	43.7
C6288[a]	33	971	32	934	3.8	75.3
C7552[a]	12	479	11	479	0.0	101.7
Total		4,282		4,070	5.0	
Average					6.5	

Ranking: Area-flow based heuristic. [a]Depth reduced.
IM: # of LUTs after IMap.
+ILR: # of LUTs after area-ILR.
% Red: Reduction percentage. **mins**: Runtime in minutes.

The number of iterations for each IMap run is 20. The LUT size K is set to be both 4 and 6. The experimental result is shown in Tables 5.11 and 5.12, respectively.

With $K = 4$, we observe significant improvement in the mapping area using logic perturbation on input circuits to IMap. The improvement is up to 21.1% (`term1`) and on average 6.5%. The runtime for the whole optimization is fast, since the efficient ranking procedure utilizing the area flow score calculation is used. The average area improvement when $K = 6$ is similar: 5.1% among the same set of benchmark. It should be noted in some cases the mapping depth is reduced as well (like `C3540` with $K = 6$).

The latest introduction of lossless synthesis by ABC allows strong depth and area improvement in FPGA technology mapping. The network is first optimized by AIG rewriting, and snapshots are made to give choices in the network (Mishchenko et al. 2007). We believe AIG rewriting and rewiring are orthogonal to each other, and rewiring can readily further improve results optimized by AIG rewriting. Thus we applied area-ILR to the mapping results by ABC. The recent `imfs` and `lutpack` routines were used to map the benchmarks with $K = 6$.

The experimental result is shown in Table 5.13, where we see the depth and area reduction is 5.2% and 5.5%, respectively, on average. The depth reduction is higher than in the previous experiments because of the optimization power given by the choices in ABC. The combination of network choices and rewiring is powerful in both depth and area reduction, provided that the combinational portion of the circuit is large enough for optimization. We would further study this direction in the future.

Table 5.12 Result of Area-ILR over IMap ($K = 6$)

	IM		IM+ILR			
Circuit	Dep.	#	Dep.	#	% Red.	mins
alu2	8	97	8	89	8.2	35.0
alu4	9	190	9	172	9.5	147.8
apex6[a]	5	166	4	161	3.0	16.1
duke2	4	103	4	101	1.9	13.4
misex3	6	125	6	111	11.2	36.5
rot	6	183	6	175	4.4	16.9
term1	4	42	4	38	9.5	4.4
x3	4	172	4	167	2.9	11.8
C1355	4	62	4	62	0.0	86.7
C1908	6	88	6	79	10.2	31.3
C2670	6	121	6	113	6.6	217.7
C3540[a]	9	250	8	255	-2.0	495.2
C5315	6	298	6	291	2.3	717.7
C6288	21	644	21	593	7.9	1,139.3
C7552[a]	8	334	7	330	1.2	3,582.4
Total		2,875		2,737	4.8	
Average					5.1	

Ranking: Area-flow based heuristic. [a]Depth reduced.
IM: # of LUTs after IMap.
+**ILR**: # of LUTs after Area-ILR.
% Red: reduction percentage. **Mins**: runtime in minutes.

We obtain 130 industrial benchmarks from an FPGA company. Our proposed area-ILR is experimented on all 130 benchmarks with both combinational and sequential circuits. The benchmarks were synthesized using the Synopsys Design Compiler and then decomposed into two-input gate networks using SIS. DAOMap is used as the technology mapping algorithms throughout this experiment, where area-ILR successfully reduced the depth or the area of 90 benchmarks. The performance of area-ILR on these responsive benchmarks is shown in Table 5.14, with benchmarks with the depth reduction shown first.

On average, it produces 21.1% depth reduction in 26 out of 130 benchmarks (20%). Despite the area penalty in some cases, the average area reduction is 3%. The depth reduction usually allows the mapped circuit to have better delay performance after place and route, as explained in the following subsection. For the remaining benchmarks (64/130, 49.2%), while the depth is maintained the area reduction is 7% on average. This shows that our technique is effective in cutting down the area when the circuit is responsive: the room for area improvement is large with respect to its initial structure. A note on the runtime: since we can only obtain executables of technology mappers, over 90% of the runtime is spent in mapping instead of searching for alternative wires. An obvious runtime optimization is to carry out incremental remapping after local transformation, which is feasible when the source code of the mapper is available.

To analyze how much real area and delay improvement can be achieved by our area-ILR scheme, we use TVPR (Marquardt et al. 2000) to place and route the optimized circuits

Table 5.13 Result of Area-ILR over imfs + lutpack in ABC($K = 6$)

	IMFS+LP		IMFS+LP+ILR		
Circuit	Dep.	#	Dep.	#	% Red.
alu4	5	413	5	384	7.0
apex2	6	449	5	496	−10.5
apex4	5	571	5	554	3.0
bigkey	3	471	2	467	0.9
clma	8	1,654	8	1,483	10.3
des	4	718	4	552	23.1
diffeq	7	532	7	511	4.0
dsip	3	689	3	689	0.0
ex1010	5	704	5	681	3.3
ex5p	3	153	3	130	15.0
elliptic	6	324	6	314	3.1
frisc	12	1,707	11	1,700	0.4
i10	8	541	7	531	1.9
pdc	6	1,052	6	932	11.4
misex3	5	271	5	214	21.3
s38417	6	2,418	6	2,353	2.7
s38584.1	6	2,253	6	2,200	2.4
seq	5	528	4	543	−2.8
spla	6	981	6	855	12.8
tseng	7	627	6	627	0.0
Total	116	17,056	110	16,216	4.9
Average					5.5

Ranking: Topological order.
IMFS+LP: # of LUTs after imfs and lutpack in ABC.
+**ILR**: # of LUTs after area-ILR.
% **Red**: reduction percentage.

obtained by our first experiments with ILR and DAOMap in $K = 4$. We measured the channel width and critical path delay reduction. The result is shown in Table 5.15.

In this experiment, we try to fit the mapped circuit into the smallest square FPGA device. This would clearly create some empty LUTs. Therefore we also present the number of slots (used and unused LUTs) for comparison. We perform timing-driven placement and routing. The routing area and critical path delay are recorded and presented in the last two sections of the table.

One important observation from this experiment is that an improvement from area minimization may not imply a reduction in the routing area. In some benchmarks like *C1908* and *C2670*, an extra channel is required because of the increased number of interconnects between LUTs. For the same reason, we can observe penalty in the critical path delay after strong area minimization in the mapping stage. As we always use the smallest possible device, such penalty is expected in highly congested FPGA design.

The above analysis reveals that despite improvement in routing area due the reduced area in the mapping stage, the critical path delay can be worse as a result more interconnects before

Table 5.14 Result of Area-ILR on Commercial Benchmarks ($K = 4$)

	DM		DM+ILR		Dep.	#
Circuit	Dep.	#	Dep.	#	% Red.	% Red.
IND035	3	17	2	16	33.33	5.88
IND024	3	19	2	22	33.33	−15.79
IND049	3	23	2	20	33.33	13.04
IND053	3	30	2	28	33.33	6.67
IND080	7	74	5	72	28.57	2.7
IND007	4	11	3	12	25	−9.09
IND012	4	18	3	19	25	−5.56
IND040	4	39	3	42	25	−7.69
IND037	4	46	3	46	25	0
IND070	4	55	3	54	25	1.82
IND100	4	75	3	72	25	4
IND102	4	138	3	143	25	−3.62
IND113	4	166	3	165	25	0.6
IND044	5	42	4	43	20	−2.38
IND031	15	46	12	41	20	10.87
IND098	5	72	4	67	20	6.94
IND084	6	80	5	72	16.67	10
IND087	6	93	5	78	16.67	16.13
IND126	6	284	5	263	16.67	7.39
IND082	7	81	6	77	14.29	4.94
IND114	8	298	7	273	12.5	8.39
IND115	8	424	7	407	12.5	4.01
IND096	10	236	9	212	10	10.17
IND106	10	399	9	364	10	8.77
IND101	11	199	10	194	9.09	2.51
IND109	12	125	11	127	8.33	−1.6
				Average$^{\alpha}$	21.10	3.04
IND003	2	5	2	4	0	20
IND005	2	10	2	9	0	10
IND022	3	12	3	11	0	8.33
IND015	3	13	3	12	0	7.69
IND010	4	14	4	12	0	14.29
IND008	4	15	4	13	0	13.33
IND014	3	21	3	17	0	19.05
IND017	6	21	6	19	0	9.52
IND028	3	22	3	20	0	9.09
IND033	3	23	3	22	0	4.35
IND051	2	25	2	24	0	4
IND032	3	26	3	25	0	3.85
IND061	3	28	3	25	0	10.71
IND020	4	30	4	29	0	3.33
IND021	4	30	4	28	0	6.67
IND060	3	30	3	29	0	3.33
IND039	3	31	3	28	0	9.68
IND034	3	32	3	30	0	6.25
IND046	4	32	4	29	0	9.38
IND059	3	32	3	24	0	25
IND025	7	36	7	30	0	16.67

(continued overleaf)

Table 5.14 (*continued*)

Circuit	DM		DM+ILR		Dep.	#
	Dep.	#	Dep.	#	% Red.	% Red.
IND058	3	36	3	31	0	13.89
IND067	3	38	3	35	0	7.89
IND066	4	44	4	42	0	4.55
IND030	3	46	3	42	0	8.7
IND076	3	48	3	43	0	10.42
IND047	6	49	6	43	0	12.24
IND071	3	49	3	47	0	4.08
IND068	3	52	3	49	0	5.77
IND075	3	60	3	55	0	8.33
IND108	3	60	3	59	0	1.67
IND074	3	61	3	54	0	11.48
IND093	3	64	3	63	0	1.56
IND065	16	65	16	63	0	3.08
IND064	2	69	2	68	0	1.45
IND027	2	80	2	48	0	40
IND086	4	82	4	80	0	2.44
IND089	3	89	3	85	0	4.49
IND088	4	90	4	84	0	6.67
IND090	4	90	4	84	0	6.67
IND069	6	92	6	87	0	5.43
IND077	8	95	8	87	0	8.42
IND078	8	95	8	87	0	8.42
IND050	8	116	8	110	0	5.17
IND112	4	119	4	114	0	4.2
IND095	32	129	32	127	0	1.55
IND104	3	131	3	124	0	5.34
IND094	19	135	19	133	0	1.48
IND097	3	151	3	142	0	5.96
IND105	21	171	21	169	0	1.17
IND110	3	171	3	168	0	1.75
IND111	3	184	3	179	0	2.72
IND099	8	222	8	213	0	4.05
IND116	12	240	12	237	0	1.25
IND117	3	265	3	262	0	1.13
IND103	7	266	7	258	0	3.01
IND118	3	266	3	263	0	1.13
IND131	3	310	3	299	0	3.55
IND129	3	311	3	307	0	1.29
IND120	4	343	4	334	0	2.62
IND121	3	357	3	344	0	3.64
IND123	4	394	4	383	0	2.79
IND122	4	399	4	387	0	3.01
IND125	10	638	10	606	0	5.02
					Average$^{\beta}$	7.09
					Overall$^{\beta}$	5.89

α: Only benchmarks with depth reduction are included.
β: Only benchmarks with no depth reduction are included.
DM: # of LUTs after DAOMap.
+**ILR**: # of LUTs after area-ILR.
% Red: Reduction percentage.

Table 5.15 Effect of area-ILR on VPR Delay performance ($K = 4$)

Circuit	#LUTs			#Slots		
	DM	+ILR	% Red.	DM	+ILR	% Red.
alu2	158	130	7.59	169	144	14.79
alu4	280	239	14.64	289	256	11.42
apex6	240	220	8.33	900	900	0
duke2	153	135	11.76	169	144	14.79
misex3	223	175	21.52	225	196	12.89
rot	256	240	6.25	961	961	0
term1	70	59	15.71	81	64	20.99
x3	243	224	7.82	900	900	0
C1355	80	78	2.50	100	100	0
C1908	133	122	8.27	144	144	0
C2670	275	274	0.36	2,209	2,209	0
C3540	352	323	8.24	361	324	10.25
C5315	543	533	1.84	1,444	1,444	0
C6288	979	649	33.71	1,024	676	33.98
C7552	544	488	10.29	1,600	1,600	0
Average			10.59			7.94

Circuit	Routing area (e+06)			Critical path delay (e−08)		
	DM	+ILR	% Red.	DM	+ILR	% Red.
alu2	0.32	0.27	14.42	5.00	5.47	−9.4
alu4	0.71	0.63	11.27	7.21	7.62	−5.69
apex6	1.66	1.40	15.69	3.97	4.32	−8.82
duke2	0.32	0.32	1.04	3.77	3.09	18.04
misex3	0.49	0.43	12.24	4.44	5.09	−14.64
rot	2.05	1.77	13.66	5.42	5.17	4.61
term1	0.11	0.11	0	2.29	2.61	−13.97
x3	1.40	1.40	0	3.80	4.00	−5.26
C1355	1.93	1.93	0	3.41	3.29	3.52
C1908	0.27	0.32	−15.64	4.99	5.03	−0.80
C2670	4.03	4.67	−15.88	6.13	5.87	4.24
C3540	0.89	0.9	−1.12	6.47	6.64	−2.63
C5315	3.06	3.06	0	6.58	6.82	−3.65
C6288	1.88	1.05	44.05	14.22	13.50	5.06
C7552	2.93	2.93	0	10.09	9.75	3.37
Average			5.32			−1.73

DM: # with DAOMap only. +**ILR**: # with area-ILR.
% **Red**: Reduction percentage.

Table 5.16 Effect of area-ILR on delay performance in commercial FPGAs ($K = 6$)

	imfs + lutpack					imfs + lutpack + ILR					Reduction %				
Circuit	#net	d	d_{10}	d_{pp}	d_{pp10}	#net	d	d_{10}	d_{pp}	d_{pp10}	#net	d	d_{10}	d_{pp}	d_{pp10}
alu4	2,181	1.62	3.52	8.64	10.38	1,901	1.53	3.29	7.92	9.93	12.9	5.63	6.62	8.33	4.34
apex2	2,472	1.33	2.89	9.39	10.31	2,480	1.23	2.67	8.81	9.61	−0.3	7.58	7.41	6.18	6.79
apex4	3,116	1.48	3.29	10.14	11.91	2,932	1.33	2.84	9.85	11.13	5.9	10.65	13.49	2.86	6.55
ex5p	788	1.37	2.41	6.34	8.39	676	1.32	2.26	5.86	7.67	14.2	3.57	6.18	7.57	8.58
pdc	5,366	1.41	3.09	8.19	12.06	4,917	1.31	2.73	7.84	11.41	8.4	7.18	11.47	4.27	5.39
misex3	1,324	1.14	2.24	7.44	8.66	1,128	1.30	2.44	7.30	9.12	14.8	−14.6	−9.07	1.88	−5.31
seq	2,824	1.67	3.58	8.42	11.41	2,833	1.61	3.24	7.68	10.39	−0.3	3.24	9.31	8.79	8.94
spla	4,758	1.53	3.51	8.05	13.08	4,581	1.36	3.13	7.74	11.5	3.7	11.13	10.92	3.85	12.08
Average											7.4	4.3	7.04	5.47	5.92

imfs + lutpack: Mapping with imfs and lutpack in ABC only.
+ **ILR**: with area-ILR.
d: Average connection delay. d_{10}: Average connection delay for worse 10 nets.
d_{pp}: Average pad-to-pad delay reported by static timing analysis in Xilinx ISE.
d_{pp10}: Average pad-to-pad delay of worse 10 nets reported by static timing analysis in Xilinx ISE.

well-packed logic modules. To further improve delay performance, we suggest to link the mapping, placement, and routing stages together, using rewiring as the perturbation technique. A study of the combined approach is presented in later sections.

VPR tries to use minimum number of channels to route the nets for a circuit placed closely in the chip. This is likely to create congestions and finally worsen the delay performance in our results. To study the effect of area-ILR on delay performance with a fixed number of routing channels, we also tried to place and route our optimized circuit using a commercial FPGA chip with Xilinx Foundation ISE.

In this experiment, we pick $K = 6$ and target our benchmarks on **Xilinx Virtex-5 FPGA**, which uses six LUTs inside their logic blocks. The benchmarks were converted to VHDL descriptions of LUT netlist (using the LUT primitive) and were placed and routed using Xilinx Foundation ISE version 9.1i. Routines `imfs` and `lutpack` in ABC were used as the mapping algorithm.

Based on the result shown in Table 5.16, we can see that the area reduction by area-ILR can also improve delay performance by reducing 7.4% number of nets to be routed. The average delay reported by the placer is reduced by 4.3%, and for the 10 most critical nets the average delay is reduced significantly (7%). The average pad-to-pad delay reported by the static time analysis is reduced by 5.5%, while the reduction is nearly the same for the 10 more critical nets.

The result has demonstrated that, in general, area minimization can improve the delay performance as a result of the smaller number of nets to be routed globally. Therefore it is justified to spend a little more runtime in order to get both area and delay performance.

Another experiment combines an existing FPGA synthesis system, and we conducted experiments on area minimization with the BDS-pga synthesis result. BDS-pga (Vemuri et al. 2002) is an LUT-based FPGA synthesis system based on BDDs. We first synthesized the benchmarks with BDS-pga (command: `bds -useloc -ethred 10 -sharing`) and then processed with the following procedures: (i) DAOMap: map into 4-LUT architecture

Table 5.17 Result of area-ILR on circuits synthesized by BDS-pga ($K = 4$)

	BDS+DM		BDS+DM+ILR			
Circuit	Dep.	#	Dep.	#	% Red.	mins
alu2[a]	6	76	5	65	14.5	4.9
alu4[a]	18	298	17	260	12.8	62.2
apex6	8	262	8	240	8.4	7.9
duke2	6	239	6	209	12.6	63.2
misex3[a]	9	277	8	209	24.5	68.3
rot[a]	10	244	9	225	7.8	3.1
term1[a]	9	124	8	101	18.5	2.0
x3	6	211	6	204	3.3	5.1
C1355	6	80	6	80	0.0	0.6
C1908	9	134	9	124	7.5	1.6
C2670[a]	10	197	8	174	11.7	5.7
C3540[a]	15	425	14	363	14.6	42.7
C5315	13	495	13	460	7.1	30.4
C6288[a]	34	1,055	33	863	18.2	214.3
C7552	17	658	17	565	14.1	101.9
Total		4,775		4,142	13.4	
Average					11.7	

Ranking: Topological order. [a]Depth reduced.
BDS: Initial logic synthesis result from BDS.
DM: # of LUTs after DAOMap.
+**ILR**: # of LUTs after area-ILR.
% **Red**: Reduction percentage.

and (ii) area-ILR: optimize the mapping area (additionally depth in some cases as well). As BDS-pga produces circuits with complex gates (XOR, XNOR gates), we first decomposed the complex gates into simple gates before we executed area-ILR. The depth and area comparisons are shown in Table 5.17.

Area-ILR reduces, on average, 11.7% of mapping area on BDS-pga results. Moreover, in eight benchmarks the mapping depth is reduced by 1. It is quite surprising that it can even further optimize the excellent synthesis results from BDS-pga; for example, the benchmark *alu2* is resynthesized with area-ILR to give 65 LUTs and depth of 5 when mapped to 4-LUTs. This shows that our system can also augment the existing FPGA synthesis systems by providing extra logic resynthesis, which is not possible in the original scheme.

5.4.4 Conclusion

As proven by the extensive experiments, there is significant room for area and depth improvement in LUT-based FPGA technology mapping when logic perturbation is used in the proposed approach. We name the framework an ILR, which can be executed in either the depth-oriented or area-oriented mode.

The depth-oriented scheme in ILR further improves mapping depth using rewiring. Experimental results show that our technique coupled with the proposed area minimization produces 11.3% depth and 6.1% area improvement on depth-optimal mapping solutions.

The area-oriented mode in ILR explores logic perturbation for area reduction in terms of the number of LUTs. The method transforms the network and makes it more suitable for the chosen technology mapping algorithms. Depth optimality is maintained, and sometimes improved, during the area optimization process. In experiments, ILR improves all tested well-known algorithms significantly: 33.7% (avg. 11.4%) versus DAOMap ($K = 4$); 21.1% (avg. 6.5%) versus IMap ($K = 4$); 23.1% (avg. 5.5%) versus imfs and lutpack in ABC ($K = 6$); and 24.5% (avg. 11.7%) versus BDS-pga ($K = 4$). Delay performance is enhanced with the area minimization. In our experiments with Xilinx Virtex-5 FPGAs, the average connection delay and pad-to-pad delay are reduced by 4.3% and 5.5%, respectively.

ILR is capable of collaborating with any technology mapping algorithm in any practical LUT sizes. Our extensive experimental results have shown that with logic perturbation we can improve both the depth and area of technology mapping beyond the purely graph-model-based optimization limit imposed by conventional technology mapping algorithms. As a next step, we will further study the application of rewiring algorithms on improving the place and route stage of the FPGA design automation and investigate how to introduce choices by rewiring.

5.5 FPGA Postlayout Routing Optimization

Conventional FPGAs use LUT-based logic blocks to implement functions. While SPFD-based rewiring algorithms are natural to use with circuits implemented on FPGAs, ATPG-based rewiring algorithms can also be suitably modified or adopted for LUT-based FPGA architecture. In this section, we will discuss various applications of rewiring algorithms in optimizing routing in FPGAs. The basic idea is to replace a wire or net with a bad cost (e.g., long wirelength or long delay) with an alternative wire/net with a better cost. Let us briefly review the conventional FPGA EDA flow, as shown in Figure 5.23.

A design is synthesized from hardware descriptive language (HDL) (e.g., VHDL, Verilog) to a gate-level representation that can be optimized by common technology-independent logic synthesis. The circuit is then mapped for the chosen technology, say a k-LUT architecture. Several LUTs will be packed together to fit in a logic block, and a netlist is generated. Two main layout steps are followed: placement assigns physical locations for every logic block while minimizing cost function like wirelength or routability; routing assigns routing resources to every net and keeps the channel width and the critical path delay as low as possible. The whole flow is concluded with the generation of the FPGA program file.

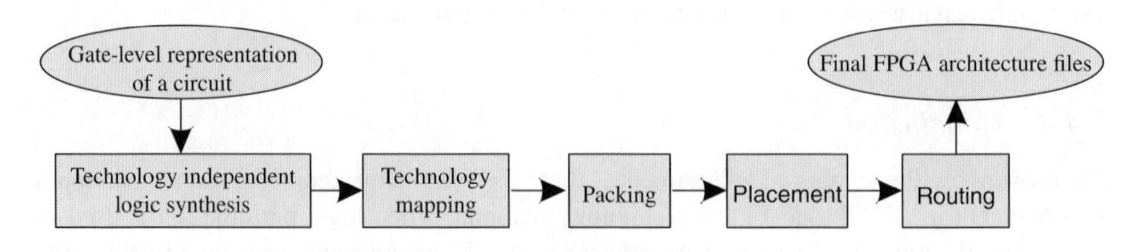

Figure 5.23 Conventional FPGA EDA flow

The main objective of postlayout optimization is to make use of some information from the completed placement or routing to pick target and alternative wires that finally give a better FPGA implementation. Yet, the tradeoff of keeping placement or routing intact is an important issue in this application. In addition, the logic information of a circuit is usually left aside after technology mapping. In order to facilitate the postlayout optimization, in real implementation of these techniques a rewiring engine has to be built to serve as an interface between a circuit's logic structure (gate-level representation) and its physical layout (LUT-level representation).

The first work applying the ATPG-based rewiring techniques for FPGA routability improvement was proposed in (Chang et al. 1999). In that work, rewiring was used to find alternative wires for all nets after placement, which was kept intact through the whole optimization. Therefore the alternative wire is added to an LUT having at least one unused input, or an existing wire is replaced. No new LUT will be generated, and only the logic function of the LUT with alternative wire addition will be changed. Therefore, we name this method optimization by *alternative function* in the following. A novel idea in this work is that routing priorities are assigned to the nets with more alternative wires. In other words, a router is advised to route wires with less number of alternative wires.

A succeeding work using ATPG-based rewiring has been introduced in Zhou et al. (2007a, b). It suggests an FPGA-layout-conscious and performance-targeted rewiring scheme for further improvement upon already excellent routings. New LUTs are generated as a result of alternative wire addition, in some cases. Therefore it aims to minimize the CPU overhead and avoid chip area penalty through a careful analysis. The basic tasks of this rewiring scheme include (i) finding alternative wires for target nets; (ii) filtering out alternative wires requiring addition of extra LUTs; (iii) maintaining the mapping depth after transformations; and (iv) using cost estimation to select alternative wires that can best improve FPGA delay performance. We call this method optimization by *mapping-to-routing logic rewiring*.

The third technique for postlayout optimization is proposed in Cong et al. (2002). Placement information is acquired to estimate the delay between LUTs. Path with a long delay (ε-critical path) will be rewired with alternative wires identified by an SPFD-based rewiring engine, which will only alter functions at the LUTs. Though this flow is primitive, it illustrates the relationship between rewiring ability and the optimization potential. We name this method optimization by *SPFD-based rewiring*.

5.5.1 Optimization by Alternative Functions

The first work applying the ATPG-based rewiring techniques for FPGA routability improvement was proposed in Chang et al. (1999).

The method of identifying alternative wire in the ATPG-based rewiring can be generalized to LUT-based FPGAs. The mapped LUTs are kept intact by using only LUTs with vacant input pins. The idea is to find an alternative function f' to be implemented in a LUT with original f, given that both the following conditions are satisfied: (i) the function of the circuit is unchanged by replacing f with f'; (ii) the support of f' is less than k. Such a valid function is an alternative function for a target wire. There are three main steps in finding alternative functions. First, the SMA of the target wire stuck-at fault is calculated; then a set of candidate functions is generated and all functions are guaranteed to force the target wire redundant. Lastly, the candidate functions are checked to be valid or not. But not all LUTs will be considered to be changed; only those that are dominators of the target wire and have at least one unused input

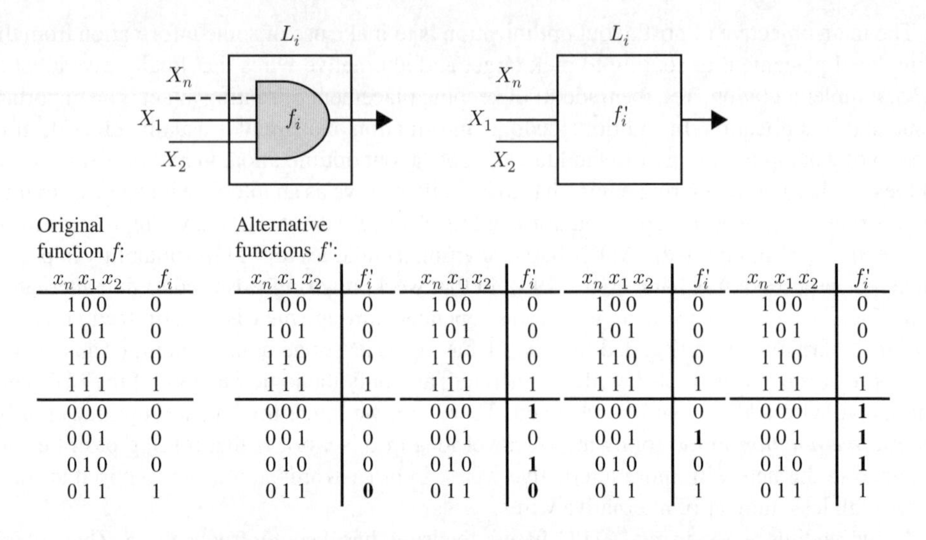

Original function f:		Alternative functions f':							
$x_n\, x_1\, x_2$	f_i	$x_n\, x_1\, x_2$	f_i'	$x_n\, x_1\, x_2$	f_i'	$x_n\, x_1\, x_2$	f_i'	$x_n\, x_1\, x_2$	f_i'
1 0 0	0	1 0 0	0	1 0 0	0	1 0 0	0	1 0 0	0
1 0 1	0	1 0 1	0	1 0 1	0	1 0 1	0	1 0 1	0
1 1 0	0	1 1 0	0	1 1 0	0	1 1 0	0	1 1 0	0
1 1 1	1	1 1 1	1	1 1 1	1	1 1 1	1	1 1 1	1
0 0 0	0	0 0 0	0	0 0 0	**1**	0 0 0	0	0 0 0	**1**
0 0 1	0	0 0 1	0	0 0 1	0	0 0 1	**1**	0 0 1	**1**
0 1 0	0	0 1 0	0	0 1 0	**1**	0 1 0	0	0 1 0	**1**
0 1 1	1	0 1 1	**0**	0 1 1	**0**	0 1 1	1	0 1 1	1

Figure 5.24 Alternative function construction using MAs

are changed to make the target wire redundant. Another option is to use an LUT with an input having a mandatory assignment. The basic idea is still to force an inconsistency in the stuck-at fault test using the new alternative wire.

Suppose we want to add a new wire to the LUT L_i. The inputs X of L_i are partitioned into (X_1, X_2), where X_1 contains all inputs in the transitive fan-out of the target wire. Given the mandatory assignments of the target wire stuck-at fault, we can identify an input x_n of L_i with a mandatory value v; if there is no such wire, then we connect a wire to L_i and name that as x_n (thus in the following x_n has an MA v). Then we try to construct a new logic function f_i' from the original function f_i: when x_n has the value $!v$, we set f_i' to be the same as f_i; otherwise we set f_i' as any function independent of X_1. One can prove that in the above process the SMA of the target wire stuck-at fault will be inconsistent.

The generation of the alternative function is illustrated in Figure 5.24. There we connect a new wire with an MA 0 to the LUT L_i. The original function of L_i is shown on the left. We can derive four alternative functions such that the stuck-at fault test of the target wire will become untestable.

Not all alternative functions are redundant, and therefore each of the generated functions has to be checked for validity. The discrepancy between the original function f and the alternative function f' can be calculated as $f \oplus f'$, and if all the cubes belongs to $f \oplus f'$ are don't cares to the circuit, then the alternative function is valid. For example, in Figure 5.24, the first alternative function has one minterm (0, 1, 1) different from the original function, while the third alternative function has one minterm (0, 0, 1) different. To verify whether a minterm or a cube is a don't care in the circuit, one can set that as mandatory assignments and see if it is possible to simultaneously justify the cube and sensitize a path from L_i to any primary output. The cube is clearly a don't care if such process fails.

The overall algorithm for finding a valid alternative function for a target wire is summarized in Algorithm 5.10.

Algorithm 5.10: Algorithm for finding alternative functions

1 **begin**
2 Calculate the SMA of target wire stuck-at fault
3 Let L_s have a mandatory assignment v (0 or 1) and L_d be a dominator, and b be a fault propating wire;
4 Suggest candidate functions f':
5 $f'(L_s! = v) = f$, $f'(L_s = v)$ = a function independent of fault propagating inputs ;
6 Test if any candidate function is valid: if all cubes that belong to $f' \oplus f$ are don't cares, the candidate function is valid;
7 **end**

In the final integration, alternative functions can be pre-computed for use by an FPGA routing tool. Routing priorities were assigned to the nets using a simple rule: nets with alternative wires were assigned lower priorities, and vice versa. Nets without alternative wires were assigned higher priorities. If a net could not be routed, it would be replaced by its alternative wire. This ranking idea roughly sounds like a wire possessing more alternative wires can be considered to have more routing flexibility, so that it can yield more alternatives in later routing stages where routing resources are less abundant. Experiments were carried out on two circuits by using the AT&T ORCA router. These two originally incompletely routed circuits were successfully routed under this straightforward scheme. The framework is illustrated in Figure 5.25.

5.5.2 *Optimization with Mapping-to-Routing Logic Rewirings*

By applying circuit rewirings, logic perturbations can be carried out by shifting logic resources from perhaps costly LUT external to cost-free LUT internal areas, or from critical to noncritical paths. A simple, while effective and low-overhead postlayout logic perturbation scheme for improving LUT-based FPGA routings without altering placements will be presented in this section. A rewiring-based logic perturbation technique is used to improve upon a timing-driven FPGA P&R tool – TVPR (Marquardt et al. 2000). Compared with the already high-quality pure TVPR results, the approach reduces critical path delay by up to 31.74% (avg. 11%) without disturbing the placement or sacrificing chip areas, where only 4% of the nets are perturbed in our scheme. The complexity of our algorithm is linear in the total number of nets of the circuit. The experimental results show that the CPU time used by the rewiring engine is only 5% of the total time consumed by the placement and routing of TVPR. Logic perturbations were also performed to improve both the technology mapping and routing to investigate the effectiveness. The results show that the best technology mapping does not always lead to the best final routing, which seems to suggest that an ideal FPGA EDA flow should consider more on tradeoffs between different stages.

In most conventional physical design tools (e.g., partitioning, placement, and routing), due to the missing of logic view information, any powerful logic-synthesis-based optimization technique is not applicable. Even with the availability of the logic view, the crucial routing performance problem is still uneasy because most of today's logic synthesis techniques, instead of being wiring-conscious are cell-conscious, which are less sensitive to today's dominating performance bottleneck – wiring delay. A layout-conscious rewiring scheme linking logic

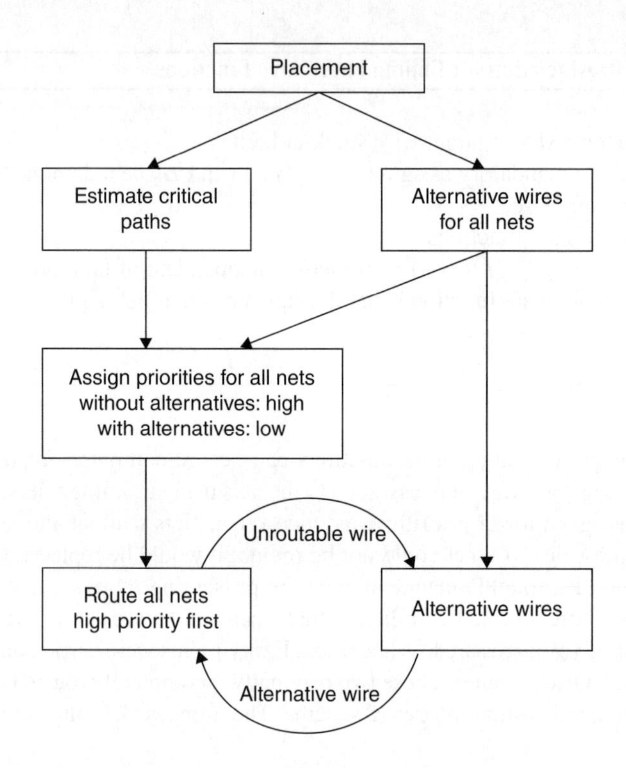

Figure 5.25 Layout-driven synthesis using alternative functions

view information with the physical layout information for improvement on performance and routability for LUT-based FPGAs is clearly beneficial. This logic-based technique clearly can be quite useful for FPGA physical synthesis stages, where quite often wiring is the main target of optimizations. Thus, rewiring can be considered as a technique binding the originally loose gap between the logical and physical optimization stages in current EDA flows.

On FPGAs, a certain structural property unique to LUTs seems to be contributive to this technique too. After technology mapping, a wire in the subject circuit will become either internal (source gate and sink gate are in the same configurable logic block (CLB)) or external (source gate and sink gate are in different CLBs). The external wires form the netlist connecting CLBs. For example, in Figure 5.26, $G1 \rightarrow G2$ could be a performance-costly external wire, and $G2 \rightarrow G3$ is an internal wire, which is virtually free of extra cost when implemented inside an LUT. Obviously, we can have a pure gain on replacing an external wire by its internal alternative wire. This is clearly some aspect not utilized well before.

Most of today's technology mapping techniques are targeted at reducing the number of LUTs and mapping depth, but without much consideration on the potential negative impact on later routings. Though a mapping depth optimized in the technology mapping step can be used as a rough objective for circuit delay reduction, this estimation can still deviate considerably from the reality after placement and routing. That is, a net may have a long routing path in the FPGA, although its source is only one level away from its sink(s). For example, Figure 5.27 shows a Boolean network and its layout after placement and routing, where $G1 \rightarrow G5$ is the

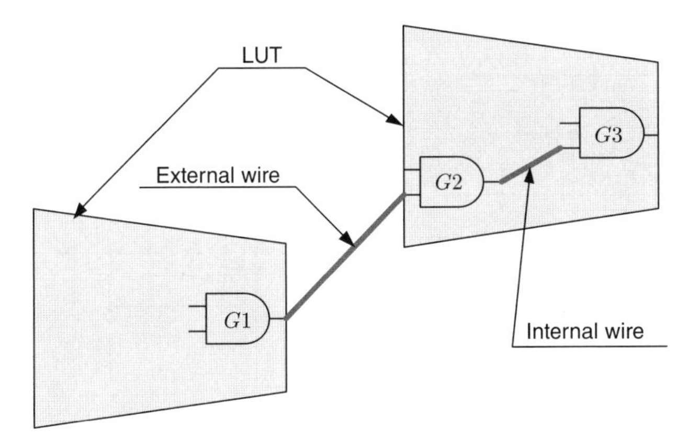

Figure 5.26 Example of external/internal wires

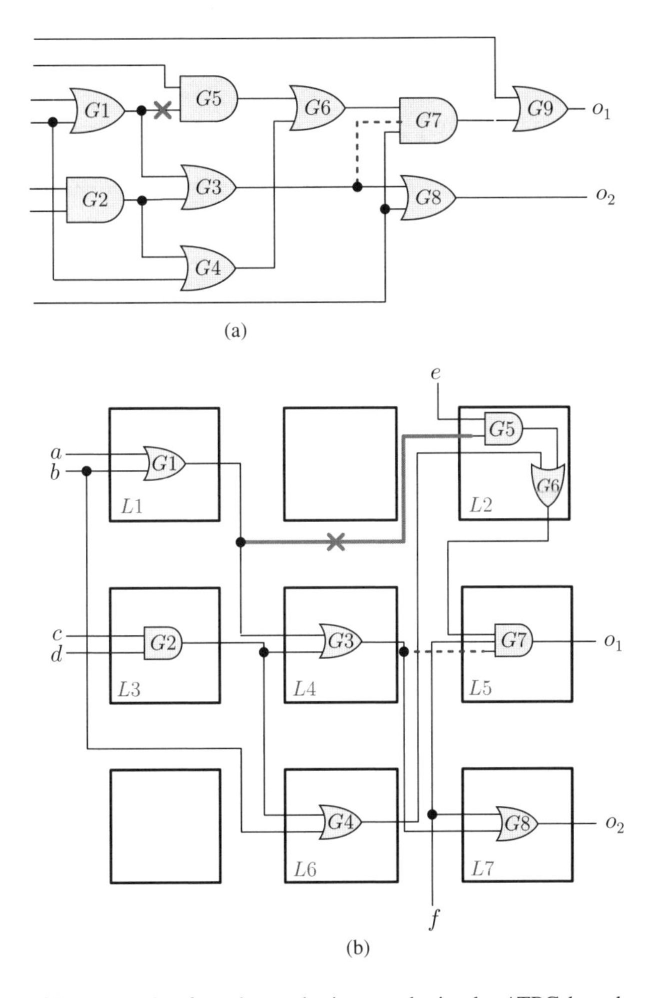

Figure 5.27 Example of postlayout logic perturbation by ATPG-based rewiring

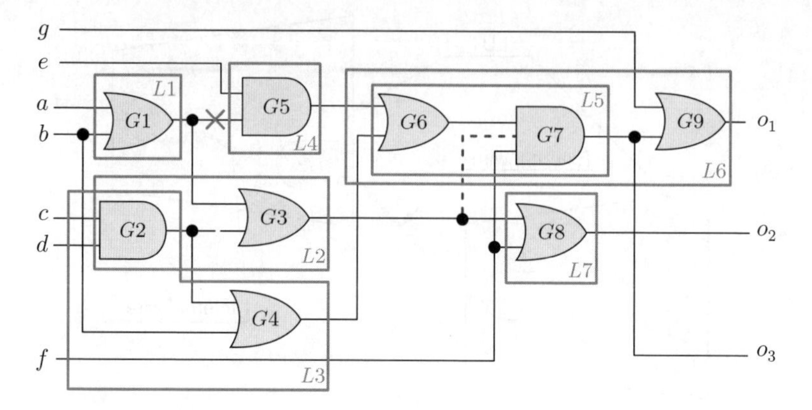

Figure 5.28 Example of multiple-wire addition

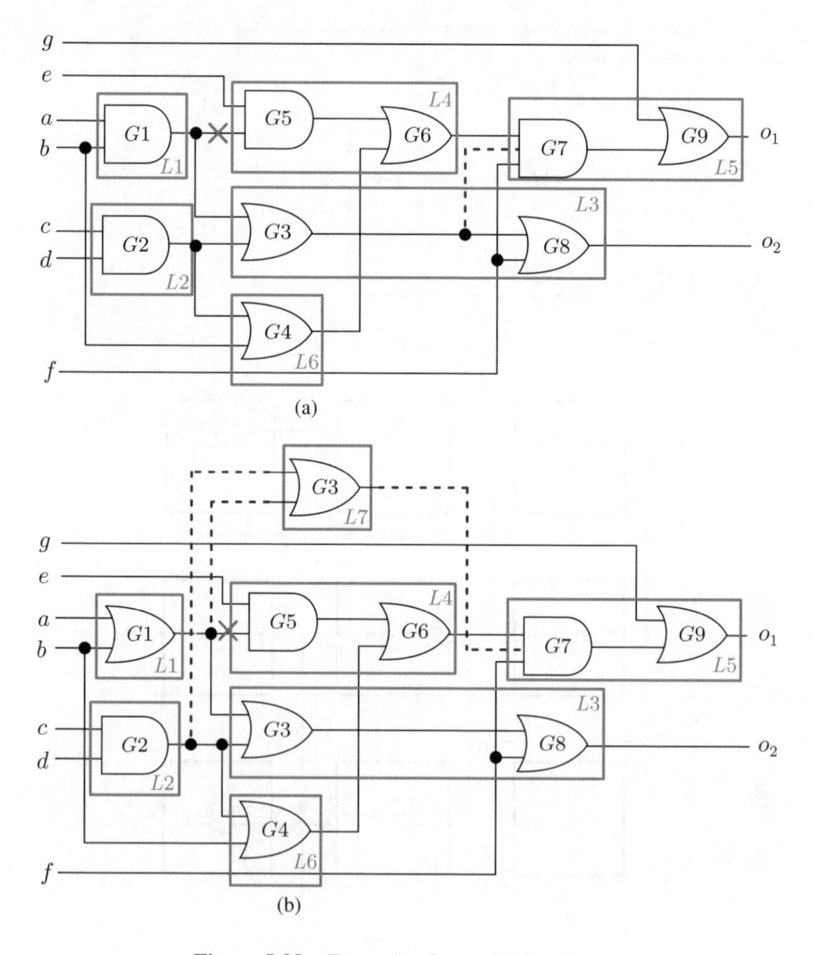

(a)

(b)

Figure 5.29 Example of extra LUT addition

target wire and $G3 \rightarrow G7$ is its corresponding alternative wire. Though the rewiring transformation does not change the number of LUTs and mapping depth, the net $L1 \rightarrow L2$ is replaced by a much shorter $L4 \rightarrow L5$. Consequently, routing becomes easier, and circuit delay is reduced because of fewer switches that the net passes through. As demonstrated by this simple example, a wiring-conscious transformation tool wisely bridging this logical–physical gap in our EDA flow would be useful. Motivated by this idea, we develop our rewiring engine to assist this postlayout logic perturbation. Through a generalized library interface, the engine can be easily integrated with any FPGA router for circuit performance optimization.

Before navigating the details of the rewiring engine, we look at the difference when rewiring is applied for technology-independent logic optimization and FPGA routing. As technology mapping forces some gates to be duplicated in several LUTs, multiple-wire addition and removal would happen. For example, in Figure 5.28, $G3 \rightarrow G7$ is added to make $G1 \rightarrow G5$ redundant and removable. $G7$ is duplicated in $L5$ and $L6$; so two wires, $L2 \rightarrow L5$ and $L2 \rightarrow L6$, are to be added for this transformation.

Based on the structural relationship between subject circuits and mapped circuits, we develop a set of rules to identify alternative candidates that can be applied for a transformation without disturbing the placement. A series of strategies are used to select good candidates for transformations.

When an alternative wire is added to the mapped circuit, new LUTs may be required to maintain the logic equivalence. For example, in Figure 5.29, $G3 \rightarrow G7$ is added to replace $G1 \rightarrow G5$. However, as $G3$ is not the output node (root) of $L3$, a new LUT ($L7$) with $G3$ being the output node will be generated, which may not be feasible as there might be no available space for the added LUT. In the following, we illustrate a set of conditions, as demonstrated by Figure 5.30, to identify an alternative wire that does not cause extra LUTs.

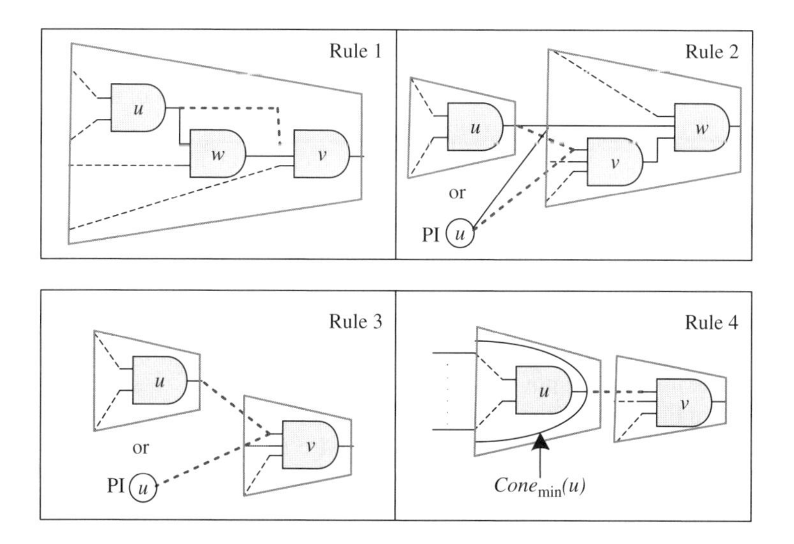

Figure 5.30 Rules for identifying alternative candidates

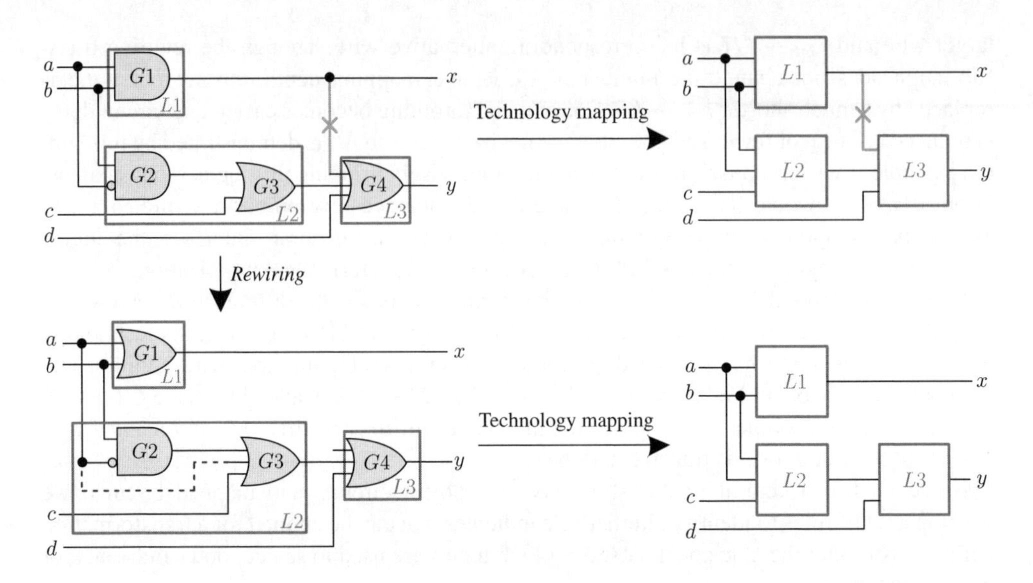

Figure 5.31 Example of an “existing” alternative wire ($K = 3$) (Rule 2)

Rule 1: $u \rightarrow v$ is internal (to a LUT).

This rule is straightforward. We only need to update the logic mapping of the LUT containing that alternative wire.

Rule 2: u is the root of an LUT (or equivalently a PI), and the LUT containing v is already taking u as an input.

If this is the case, then clearly an internal wiring branching can be freely done by logic remapping of the LUT containing v. The example in Figure 5.31 demonstrates this rule: $a \rightarrow G3$ is to be added to replace $G1 \rightarrow G4$, whereas there has already been a wire $a \rightarrow L2$ connecting a and $G2$. So we only need to update the mapping of $L2$ without adding extra LUTs or nets.

Rule 3: u is the root of an LUT (or equivalently a PI), not yet an input of the LUT containing v, and the LUT containing v has an unused pin to connect u.

As a gate may be duplicated in several LUTs in a technology mapping, there can be more than one LUT containing v. Therefore, the application of Rule 2 or Rule 3 needs to ensure that all related LUTs satisfy the requirements. Sometimes, we may need to add several wires to one LUT simultaneously (Rule 4), which requires the destination LUT to have enough free input pins for all new wires.

Rule 4: u is neither a PI nor the root of an LUT. Given that $M1$ is the input set of u's TFI cone inside the LUT containing u, and M2 is the input set of the LUT containing v, $|M1 + M2| \leq K$ (maximum input pin number of a LUT).

For example, in Figure 5.32, the TFI cone of $G6$ inside the LUT containing $G6$ only covers $G4$, $G5$, and $G6$. Generally, given an alternative wire $u \rightarrow v$, if $|M1 + M2| \leq K$, we can then duplicate the whole logic producing u with the input set $M1$, inside the LUT containing v. Thus, we do not have to introduce an extra LUT. This process is called expansion. For example, in Figure 5.33, $G3 \rightarrow G7$ is to be added to make $G1 \rightarrow G5$ redundant and removable. Considering $M1 = \{G1, G2\}$, $M2 = \{G6, f\}$, $K = 4$, and $|M1 + M2| = K$, we can then expand $L5$ by connecting the inputs of $M1$ ($G1$ and $G2$) to the duplicated logic $G3$ inside $L5$. Thus, the transformation is completed by updating the mapping of $L5$ with the connection of two new wires without any LUT addition.

After a technology mapping, some gates may be duplicated inside several LUTs, and therefore a gate can have more than one TFI cone based on different LUTs it is included in.

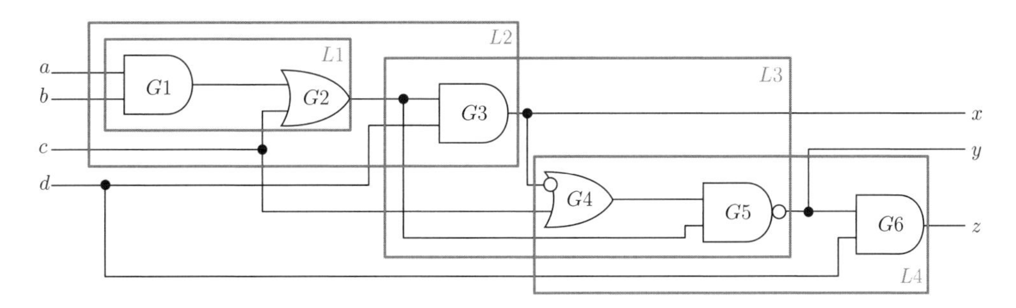

Figure 5.32 Example of gate duplication in technology mapping ($K = 4$)

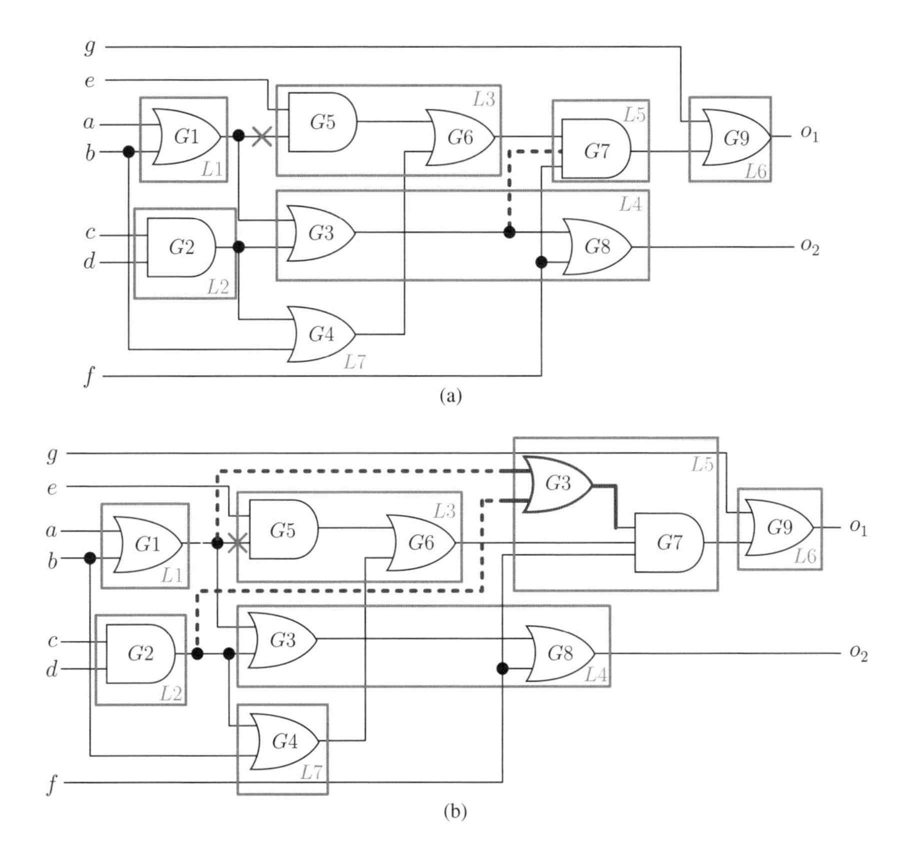

Figure 5.33 Example of destination LUT expansion ($K = 4$) (Rule 4)

Obviously, if this gate is the source node u of the alternative wire $u \to v$, then choosing this gate's smallest related input set, that is, the minimum TFI cone, might increase the chance of successfully expanding all LUTs containing v. For example, in Figure 5.32, $G4$ is duplicated in $L3$ and $L4$. Its TFI cone in $L3$, Cone3, contains $G3$ and $G4$ with input set $\{G2, c, d\}$, whereas the TFI cone of $G4$ in $L4$, Cone4, only contains $G4$ with input set $\{G3, c\}$. So the minimum TFI cone of $G4$ is $\{G3, c\}$. When an alternative wire starting from $G4$ is to be added, $G3$ and c will be connected to all LUTs containing the sink node under Rule 4.

When a new wire is added to the mapped circuit, it is important to ensure that the depth of the circuit is not increased to help avoid increasing the critical path delay. Therefore, the mapping depth has to be maintained well using the following labeling procedure. Each LUT L is assigned a label, denoted by $label(L)$, based on its topological order in the mapped circuit with primary inputs assigned label 0. Thus, the label of L is always larger than that of all its inputs. In other words, $label(L2) > label(L1)$ if $L1$ is an input of $L2$. When a new wire $L1 \to L2$ is considered, it will be taken for transformation only when $label(L1) < label(L2)$, and $label(L2)$ will keep its original label. The rewiring engine performs the label checking to eliminate all candidates that may cause any node's label increase to ensure that the mapping depth of any path is not increased after any rewiring. When several new wires are added to expand a destination LUT $L2$, we get the maximum label $l_{\max}$ from all new input LUTs. If the condition $l_{\max} < label(L2)$ is satisfied, the new wires can be accepted for transformations.

As a rewiring algorithm may find more than one feasible alternative wire for a target wire, we have to select the first candidate whose length satisfies the length constraint set by

$$LEN(AW) \leq LEN(TW) + \alpha \tag{5.8}$$

All lengths are measured using their Manhattan distance, with block spans being the length units. α is an integer specified by users, and $LEN(TW)$represents the target net length. If a candidate's length, $LEN(AW)$, is smaller than or equal to $LEN(TW) + \alpha$, it will be accepted; otherwise, the engine will go on measuring the next candidate net until one is found, or this target wire is abandoned. The parameter α is used to relax the length constraint on the candidate: if it is too small, more candidates will be filtered out including some effective ones; if it is too large, a much longer candidate, which might degrade the delay performance, will be selected. Empirically, $\alpha = 3$ provides the best results. It might appear bizarre on the first look at this formula, since it allows the alternative wire to be longer than its target wire. However, this happens to be the magical part of applying rewiring techniques. Through keeping a mildly relaxing filtering on the alternative wire lengths, a larger performance gain can be obtained by "breaking" the originally critical path delay via replacing some critical path wires by some – though might be longer – alternative wires located at noncritical paths.

The linear cost function borrowed from VPR is used to evaluate the chosen candidates. A candidate that costs more than its target net will be discarded; otherwise, the transformation will be performed. The cost function is written as

$$Cost = \sum_{i=1}^{N_{nets}} q(i) \left[\frac{bb_x(i)}{C_{av,x}(i)^{\beta}} + \frac{bb_y(i)}{C_{av,y}(i)^{\beta}} \right] \tag{5.9}$$

where N_{nets} is the total number of nets. $bb_x(i)$ and $bb_y(i)$ denote the horizontal and vertical spans of bounding box of net i. $C_{av,x}(i)$ and $C_{av,y}(i)$ indicate the average channel capacity in the horizontal and vertical directions over the bounding box. The parameter β is used to

Table 5.18 $q(i)$: Number of terminals versus net weight

# of terminals	Net weight q	# of terminals	Net weight q
1–3	1.0000	15	1.6899
4	1.0828	20	1.8924
5	1.1536	25	2.0743
6	1.2206	30	2.2334
7	1.2823	35	2.3895
8	1.3385	40	2.5356
9	1.3991	45	2.6625
10	1.4493	50	2.7933

adjust the relative cost of using narrow and wide channels. The larger the value of β, the more wiring in narrow channels is penalized relative to wiring in wider channels. Empricially, $\beta = 1$ gives placements of the highest quality. The parameter $q(i)$ is used to approximate the routing resource demand inside the bounding box. Its value depends on the number of terminals on the net i, as Table 5.18 shows. Suppose all channel capacities are the same, and the numbers of terminals on all nets are equal; then the smaller each net's bounding box, the lower the cost, and the better the placement. Based on this idea, given that all nets are two-pin nets and all channel capacities are equal, when a longer wire is replaced by a shorter wire, the cost will be reduced, and the transformation can be performed. But in practice, most nets are multipin nets, and the removal or addition of a subnet may not change the net's bounding box. Thus, the change in cost depends on the value of $q(i)$ instead. Even though the bounding box size is different after a transformation, the cost change depends not only on the bounding box but also on the parameter $q(i)$. So given a net, we select one wire from the alternative candidate set using Equation 5.8, which is simply based on the bounding box computation of a single wire, and evaluate its efficiency using Equation 5.9, which is more accurate in reflecting the relation between the layout and the netlist structure.

The logic rewiring engine can be integrated in the conventional FPGA router in the way illustrated in Figure 5.34. First, the input circuit is placed and routed without transformations. From the routing results, we obtain the channel width (W), delay of each net, and nets on the critical paths. Then the rewiring engine is activated to find alternative wires for the nets on the paths whose delays are larger than $(1 - \beta)T$, where T is the critical path delay and $\beta < 1$. A new netlist is formed after a few rewiring transformations. The original FPGA router reroutes the new netlist with channel width W. Finally, the final FPGA implementation result is obtained. For instance, we can integrate the rewiring engine in TVPR to evaluate the effectiveness of the rewiring scheme when applied for the postlayout placement-intact situation.

In most ATPG-based rewiring algorithms, the CPU time spent for a stuck-at fault test at w_t is bounded by $O(N^2)$ even when a whole chip is traversed, where N is the number of gates. However, as in the rewiring implementations, the logic implication process is usually only done over a small local range and the recursive learning depth specified by users is usually only one or two, the time complexity for finding an alternative wire for any w_t is nearly constant in practice. In the rule-based rewiring synthesis for routing delay, the number of screening rules is fixed, and the rewiring engine only processes nets on critical paths and one time only.

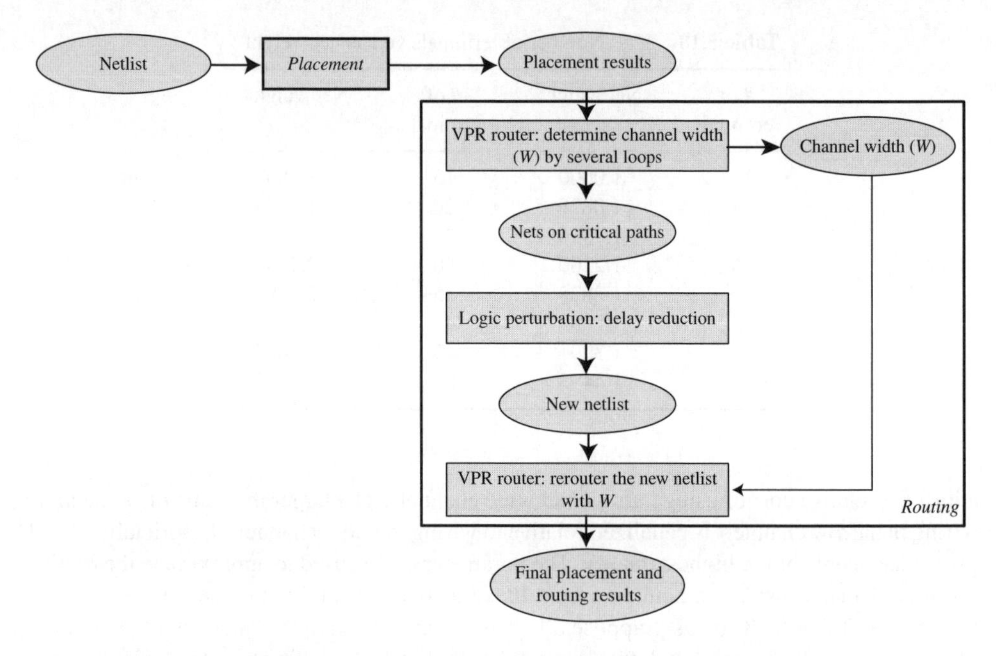

Figure 5.34 Work flow of rewiring-based FPGA router

The worst case is when all nets are on critical paths. Suppose E is the total number of nets; then the time complexity of our logic perturbation is theoretically bounded by $O(EN^2)$ but virtually near $O(E)$, as the N^2 term is degenerated to a constant in practical implementations.

Experiments on MCNC benchmark circuits were carried out to evaluate the efficiency of the ATPG-based rewiring techniques in postlayout logic perturbation for FPGA routings. All the circuits were mapped using four-input LUTs by DAOMap, and each CLB contains a single LUT. The timing-driven routing algorithm is chosen in TVPR.

Table 5.19 shows the experimental results in rewiring ability, critical path delay, and CPU time. Columns 2–4 show that 3.64% of all nets are replaced by their alternative wires for delay performance improvement. Although rewiring can find many more alternative wires, only a small part of them are useful in delay reduction. Columns 5–7 show the comparison results of critical path delay. The average delay reduction is nearly 11%, with the highest being 32% without the increase of channel width. In columns 8–10, the CPU time consumed by the engine is only 5% of the total time for TVPR's placement and routing, which is much faster than all other known approaches. All the benchmark circuits could be placed and routed within 3 minutes.

Besides routing, the ATPG-based rewiring has also been applied in technology mapping for reducing the number of CLBs. DAOMap, one of the most powerful academic tools for technology mapping, was used in this experiment for the comparison. Logic perturbation in routing was performed after that in rewiring-based technology mapping. Table 5.20 shows that by applying rewiring in both stages, though the number of logic blocks is reduced by over 8%, the delay reduction is only 3.8%. This also implies an anomaly point: *it is not necessarily true that the best technology mapping always yields the best final routing result.*

Table 5.19 Experimental results on rewiring-based FPGA router

Circuit	# *Tr.*	# Nets	Ratio (%)	Critical path delay (e−08 s) TVPR	RTVPR	Red.(%)	CPU time (s) Engine	TVPR	Ratio
5xp1	2	43	4.65	2.49	1.7	31.74	0.12	1.31	0.09
C1355	0	121	0	3.41	3.41	0	0.07	10.03	0.07
C6288	0	1,011	0	14.22	14.22	0	0.98	139.9	0.01
C880	6	180	3.33	4.11	3.48	15.33	0.19	13.81	0.01
alu2	18	168	10.71	5	4.79	4.15	2.59	30.65	0.08
apex6	0	375	0	3.97	3.97	0	2.33	101.65	0.02
comp	2	64	3.13	3.06	2.47	19.37	0.03	1.3	0.02
duke2	6	175	3.43	3.77	3.3	12.57	1.87	25.18	0.07
f51m	1	50	2	2.17	1.93	11.06	0.2	1.6	0.12
pcler8	0	65	0	1.87	1.87	0	0.02	1.34	0.01
term1	11	104	10.58	2.29	1.99	12.91	0.24	4.74	0.05
ttt2	7	88	7.95	2.49	1.81	27.11	0.14	3.44	0.04
x3	6	378	1.59	3.8	3.68	3.1	1.62	71.2	0.02
Average			3.64			10.56			0.05

[a]*Tr.*: Wire transformation. RTVPR: TVPR with rewiring engine. Red.: Reduction.

Table 5.20 Experimental results on rewiring-based technology mapping and routing

Circuit	#CLBs No RW	(TM+RW) + (RT+RW)	Red. (%)	Critical path delay (e−08 s) no RW	(TM+RW) + (RT+RW)	Red. (%)
5xp1	36	33	8.33	2.49	1.93	22.49
C1355	80	78	2.5	3.41	3.41	0
C2670	275	274	0.36	6.66	5.47	17.87
C880	120	119	0.83	3.6	3.6	0
alu2	158	130	7.59	5	5	0
b9	46	41	10.87	2.08	2.08	0
f51m	42	39	7.14	2.17	2.17	0
misex3	223	175	21.52	5.03	5.45	−8.35
pcler8	38	37	2.63	1.87	1.87	0
term1	70	59	15.71	2.29	2.29	0
ttt2	64	56	12.5	2.49	2.05	17.67
x3	243	224	7.82	3.8	3.98	−4.74
Average			8.15			3.75

RW: Rewiring. TM: Technology mapping. RT: Routing. Red.: Reduction.

Therefore, an EDA flow with more stages integrated together and a powerful logic perturbation tool is highly recommended to shift optimization resources between them for a globally best final solution.

5.5.3 Optimization by SPFD-Based Rewiring

This primitive optimization flow using SPFD-based rewiring can be carried out after placement. Since SPFD-based rewiring does not perturb the placement, a delay model can be easily defined: based on the locations of LUTs in placement, we use statistics to calculate the delay between different locations in the placement. The delay between two LUTs is estimated as the average delay between these two locations. A path is considered ε-critical if its delay is larger than $(1 - \varepsilon)D$, where D is the largest (estimated) path delay and $\varepsilon < 1$. The parameter ϵ will be increased gradually during the optimization.

Given the placement information, the SPFD-based rewiring engine traverses the circuit for a number of passes to perform rewiring on the ε-critical paths. When all transformations have been committed, a final routing is executed for the whole new netlist without redoing the placement.

In experiments, Quartus (Version II 1.0) was applied to do placement and routing. The application of SPFD-based rewiring brought about a reduction of up to 22.3% (avg. 5.1%) on critical path delay, whereas 2 of the 11 benchmark circuits became worse on delay performance. Table 5.21 shows the experimental results.

The approach suffers from a very slow runtime. According to the experiments, the placement and routing for some circuits were not completed within 8 hours because of their CPU-costly equivalence condition test for the rewiring. For the other circuits, the runtime of SPFD-ER was 12.5 times that of the SPFD-LR algorithm. The tradeoff between quality and CPU time should be clearly noted.

Table 5.21 Postlayout optimization by SPFD-ER

Circuit	Original delay	Optimized delay	Reduction (%)
C1908	15.598	15.625	−0.2
C432	24.731	22.198	10.2
alu4	18.696	16.65	10.9
apex6	8.456	8.149	3.6
dalu	11.872	11.772	0.8
example2	6.482	6.488	−0.1
term1	6.945	6.927	0.3
x1	6.778	6.711	1.0
x3	7.706	7.358	4.5
alu2	14.64	11.378	22.3
C5315	14.384	14.005	2.6
Average			5.1

5.6 Logic Synthesis for Low Power Using Clock Gating and Rewiring

Low power dissipation for digital circuits has become a major requirement in today's electronics industry. Minimizing power dissipation is now an important topic for design automation tools. Power consumption can be reduced significantly by reducing the dynamic power because dynamic power contributes more than 50% of the total power (Donno et al. 2004). Therefore, clock gating is usually adopted to reduce clock network loading as well as dynamic power dissipation for sequential circuits.

The authors of Fraer et al. (2008) and Babighian et al. (2005) focused on register-transfer level (RTL) clock gating. Their major idea is to compute the clock gating conditions based on the structural information of the given circuit. In some research such as Benini et al. (1994) and (1999), the authors addressed the issues of clock gating at the symbolic level. They proposed to compute the clock gating conditions in a circuit by analyzing the state information of the finite state machines representing the circuit. In some recent studies Hurst (2007) and Babighian et al. (2007), the authors presented an interesting idea of using simulation trace to extract the existing logic in the circuit, which can be used as clock gating conditions. The researchers in Li et al. (2003) suggested to calculate clock gating conditions by analyzing the deterministic usage patterns of the circuit block.

A clock gating condition is beneficial only if its overhead is small. A specialized rewiring-based transformation scheme (Lam et al. 2010) is discussed in this section to evaluate its effectiveness in overhead reduction.

5.6.1 Mechanism of Clock Gating

The traditional clock gating techniques work by disabling the clock inputs of some sequential components in a circuit whenever the outputs of these sequential components are unobservable or are not switching in the next clock cycle.

5.6.1.1 ODC-Based Clock Gating

In the example circuit shown in Figure 5.35(a), the circuit components $D1$ and $D2$ are flip-flops that propagate their input values in the previous clock cycle to their output in the next clock cycle during the edge of clock transition. The ODC of d^+ is b^+. This is because when $b = 1$ in the previous clock cycle, $b^+ = 1$ in the next clock cycle, and then the value of d^+ will be unobservable because the controlling value of OR gates is 1. Hence, there is in fact no need to update d^+ for the next clock cycle. The flip-flop $D1$ can be clock-gated using this clock gating condition. This is illustrated in Figure 5.35(b). Clock gating is achieved by generating the logical conjunction (AND) of b and the clock signal so that whenever $b = 1$, the resultant clock signal feeding $D1$ switches to zero.

The condition $b^+ = 1$ in this example is known as an ODC of d^+. ODCs (Damiani and De Micheli 1990) of a Boolean variable are the conditions under which the variable do not affect any of the primary outputs. They are formally defined as in Equation 5.10, where ODC_x^f represents the ODC of a variable x with respect to the function f. ODC_f is the ODC of f with respect to all the primary outputs:

$$ODC_x^f = f|_x \bar{\oplus} f|_{\bar{x}} + ODC_f \tag{5.10}$$

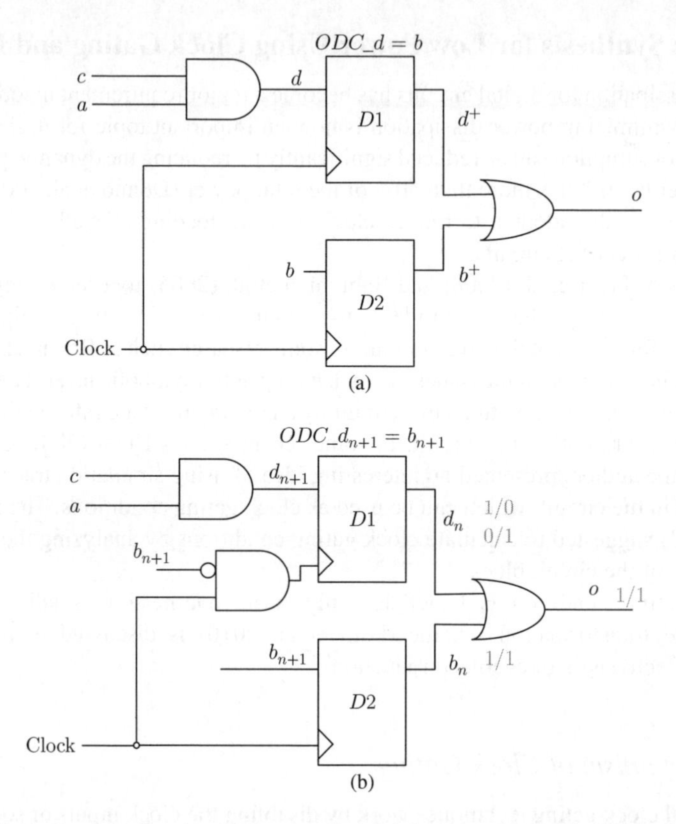

Figure 5.35 Example of clock gating implementation. (a) A sequential circuit; (b) clock-gated sequential circuit

Note that the definition is recursive. The variable x is clearly unobservable with respect to the primary outputs under the condition when the function f is itself unobservable. A variable may be the support of more than one functions. Therefore, the calculation of its ODC is defined in Equation 5.11:

$$ODC_x = \prod_{i=1}^{n} ODC_x^{f_i} \tag{5.11}$$

This is to compute the logical conjunction of the ODCs of the variable x with respect to each of the functions from f_1 to f_n that it supports. The meaning of this definition is that a variable is unobservable if it is unobservable with respect to all its fanouts.

The mechanism of ODC-based clock gating is now clear. If a variable is the inputs of some sequential components and it is unobservable in the next clock cycle, the clock of those sequential components for the next clock cycle can be disabled (Figure 5.37(b)).

As discussed in Lam et al. (2010), using Equations 5.10 and 5.11 directly to calculate clock gating conditions will produce an incorrect result. Figure 5.36 is used to illustrate the problem of incompatibility of ODCs in clock gating. In the current clock cycle, $a = 1$, $b = 1$, $a^+ = 0$, and $b^+ = 0$. Then, the value at the input of $D3$ is 0. If we apply Equations 5.10 and 5.11 to calculate the ODCs of a and b directly, we get $ODC_a = b$ and $ODC_b = a$. Under the current

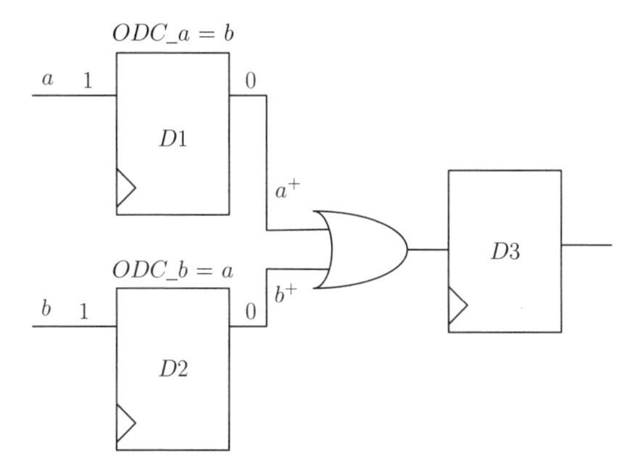

Figure 5.36 Flip-flops that are mutually unobservable

condition and ODC calculation method, the two flip-flops $D1$ and $D2$ are considered to be unobservable and can be clock-gated in the next clock cycle. However, if no clock gating is applied, the value of the input of $D3$ should be 1 instead. This problem is known as the incompatibility of ODCs (Brayton 2001, Saluja and Khatri 2004).

The error-free ODCs of $D1$ and $D2$ should be $ODC_a = b\overline{ba} = b\bar{a}$ and $ODC_b = a\overline{ba} = a\bar{b}$ respectively. However, this solution may be too restrictive because the clock gating conditions are the logical conjunction of two variables. The more complex a clock gating condition is, the more the area overhead that will be introduced. Therefore, clock gating conditions should be as simple as possible. There can be other more beneficial solutions. For instance, it may be more feasible to have $ODC_a = b$ and $ODC_b = \phi$ if the switching activity of a is higher that of b. The authors of Lam et al. (2010) proposed an algorithm to calculate ODC while taking compatibility into account in the context of clock gating. Readers may refer to their paper for more details on proper ODC calculation for clock gating.

5.6.1.2 Idle-Based Clock Gating

When the input and output values of a flip-flop are the same in the current clock cycle, the value of the flip-flop's output in the next clock cycle will not change. The flip-flop is said to be in a nonswitching state and its clock can be disabled in the next clock cycle. Let x_R be the variable in the current clock cycle and $N_R(x_R)$ be the function of register R in the next clock cycle. The clock gating conditions of a flip-flop whose output value will not switch is defined as follows:

$$NS_R = N_R(x_R)\bar{\oplus}x_R \tag{5.12}$$

An example of this kind of clock gating is shown in Figure 5.37(c).

5.6.1.3 Circuit for Clock Gating

The circuit for disabling the clock typically consists of a latch and an AND gate (for positive edge-triggered flip-flops). As shown in Figure 5.37(a), when the clock gating conditions is true,

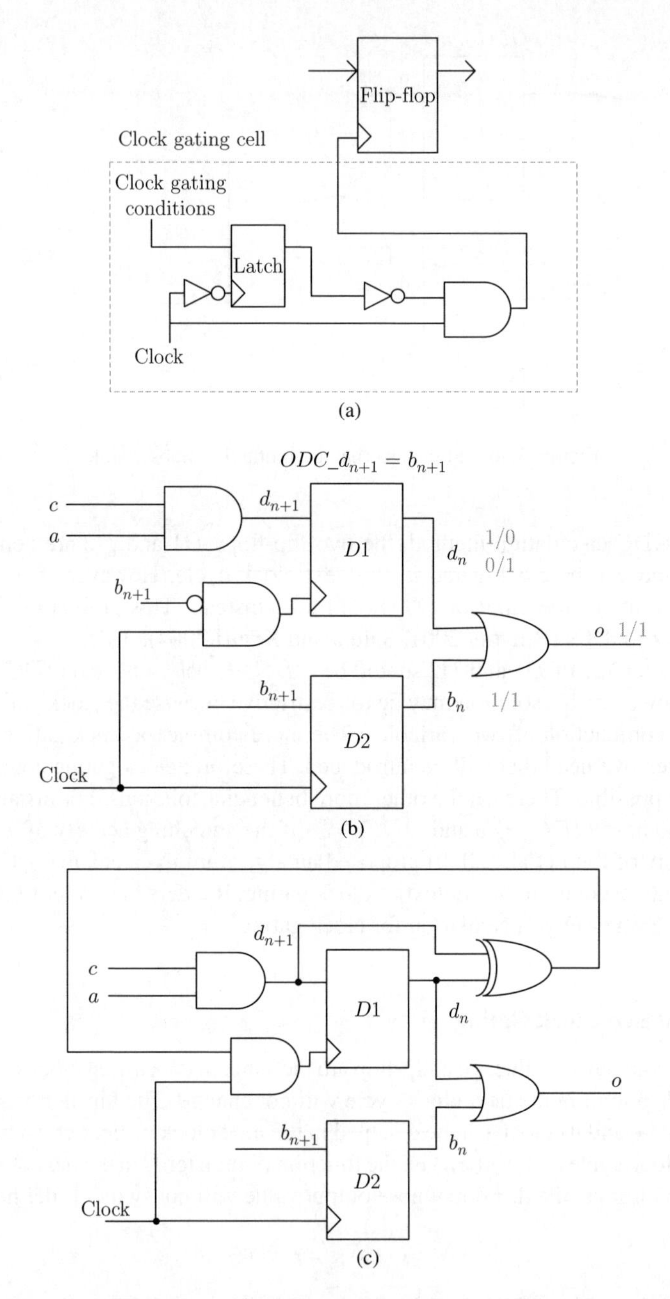

Figure 5.37 Examples of clock gating. (a) Clock gating; (b) observability don't cares based clock gating; (c) idle-based clock gating

the switching clock signal is blocked by the AND gate. The latch is used to reduce glitches. In some standard cell libraries, there are specific cells implementing the clock gating circuit.

5.6.2 Rewiring-Based Optimization

Clock gating information can be utilized to guide the restructuring of the circuit for low power. Based on the fact that the value of the output of a clock-gated flip-flop switches with relatively low frequency and hence the flip-flop can afford to drive more fanouts or loads, three types of rewiring-based logic transformations can be derived to enlarge the transitive fanout cones of those clock gated flip-flops.

5.6.2.1 Rewiring Strategy I

The transitive fanout cones of the clock-gated flip-flops are extracted for rewiring analysis. Let the set of clock-gated flip-flop be F, and let their outputs be F_o. If an alternative wire has its source in F_o and its destination not in F_o, the corresponding target wire is replaced by the alternative wire. This concept is shown in Figure 5.38.

5.6.2.2 Rewiring Strategy II

This heuristic aims to enlarge the transitive fanout cones of the clock-gated flip-flops so as to reduce the switching activity of more combinational cells. This can be achieved by replacing

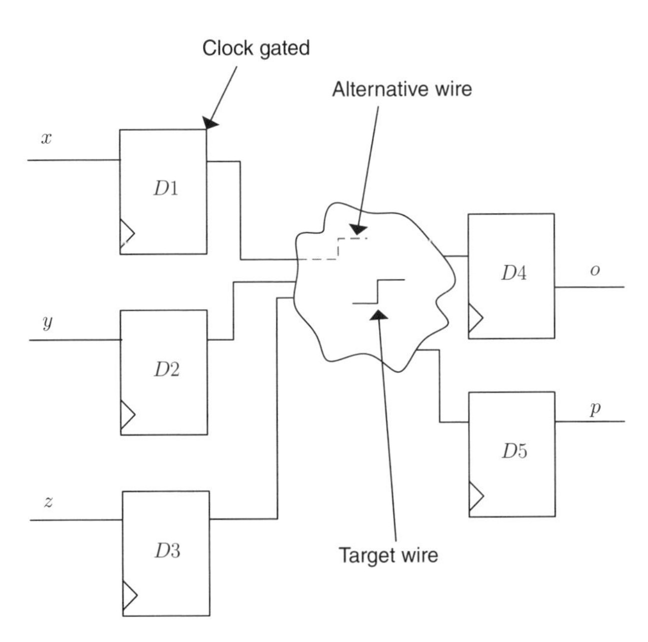

Figure 5.38 Rewiring strategy I

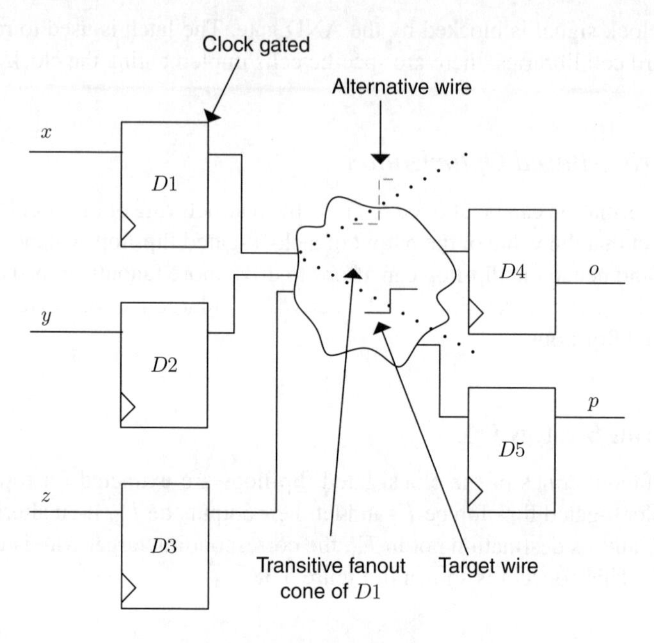

Figure 5.39 Rewiring strategy II

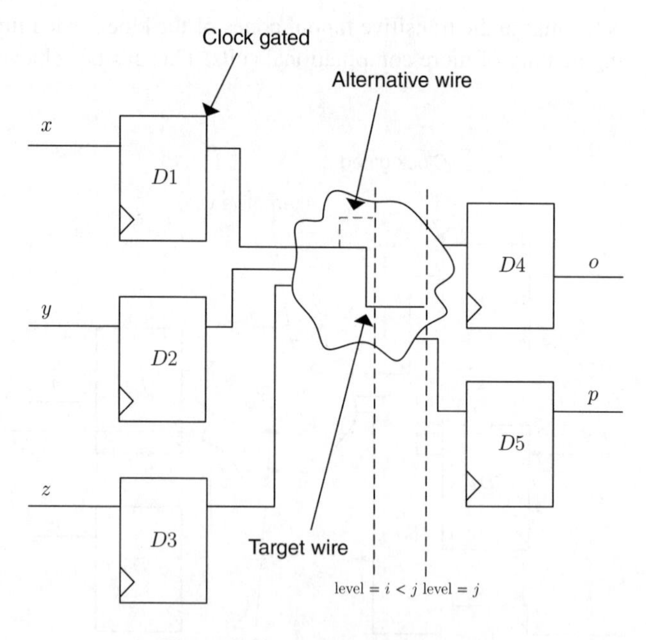

Figure 5.40 Rewiring strategy III

wires whose sources are not in the transitive fanout cones of the clock-gated flip-flops with new wires whose sources are in those transitive fanout cones. Figure 5.39 shows this idea.

5.6.2.3 Rewiring Strategy III

A signal can affect the switching frequencies of other signals by making connections with them via some additional gates. For example, a signal that switches less frequently may be connected with another signal via an AND gate to reduce the switching frequency of the latter signal. The effect of such kind of restructuring should be operative as soon as possible. Therefore, the fanout wires of the clock-gated flip-flops should be connected to gates that are closer the flip-flops, as shown in Figure 5.40.

The results of optimizing a set of benchmarks using the approach proposed in Lam et al. (2010) are listed in Tables 5.22 and 5.23.

Table 5.22 Total cell areas of the circuits

	Original	Clock-gated		Rewired	
	Area	Area	δ area (%)	Area	δ area (%)
s27	1,071.16	1,282.86	19.76	1,282.86	19.76
s526ns	8,127.32	8,527.08	4.92	7,698.34	−5.28
sse	5,373.30	5,828.59	8.47	5,042.72	−6.15
s713s	9,110.43	9,193.10	0.91	8,132.80	−10.73
s641-r	11,892.15	12,392.34	4.21	11,752.29	−1.18
s641-retime	11,892.15	12,392.34	4.21	11,739.51	−1.28
s386	5,701.43	5,866.76	2.90	5,783.28	1.44
s444	9,532.21	10,216.99	7.18	9,744.61	2.23
s526n	10,749.09	10,776.64	0.26	9,751.09	−9.28
s526	10,888.70	10,772.84	−1.06	9,673.30	−11.16
styr	19,725.20	21,952.76	11.29	19,578.56	−0.74
s382	9,559.66	10,230.64	7.02	9,840.43	2.94
s510	8,919.32	9,034.65	1.29	8,646.33	−3.06
s400	9,582.22	10,171.34	6.15	9,785.44	2.12
s641	11,792.25	11,920.85	1.09	11,049.27	−6.30
s832	10,483.42	12,770.91	21.82	12,102.11	15.44
s713	11,525.00	11,681.37	1.36	11,043.61	−4.18
s953n	16,104.45	17,208.21	6.85	15,543.70	−3.48
scf	34,200.42	35,230.65	3.01	33,827.07	−1.09
s1238	24,963.87	27,267.77	9.23	24,990.43	0.11
s1423	36,777.42	37,319.96	1.48	37,341.76	1.53
s1494	23,167.57	24,736.64	6.77	23,948.54	3.37
s1488	23,128.17	24,792.36	7.20	23,853.20	3.13
s5378	80,266.48	81,871.75	2.00	81,297.66	1.28
s13207	243,757.10	258,497.19	6.05	200,605.31	−17.70
s15850	255,331.79	274,887.34	7.66	249,506.02	−2.28
Average			5.85		−1.17

Table 5.23 Power dissipation statistics

	Com.	Seq.	Original		Clock-gated			
			D	C	D	C	δD (%)	δC (%)
s27	10	3	4.01E−06	9.59E−07	4.03E−06	7.48E−07	0.45	−21.95
s526ns	81	19	2.47E−05	6.51E−06	2.06E−05	4.34E−06	−16.89	−33.25
sse	91	4	1.15E−05	1.07E−06	1.19E−05	7.15E−07	3.39	−32.91
s713s	123	14	7.88E−06	1.46E−06	7.84E−06	1.39E−06	−0.48	−4.73
s641-r	157	18	1.02E−05	1.88E−06	9.96E−06	1.60E−06	−2.36	−14.66
s641-retime	157	18	1.02E−05	1.88E−06	9.92E−06	1.60E−06	−2.24	−14.66
s386	85	6	1.03E−05	1.60E−06	7.69E−06	6.24E−07	−25.56	−60.98
s444	98	21	2.66E−05	7.19E−06	1.49E−05	1.38E−06	−44.04	−80.84
s526n	126	21	2.83E−05	7.19E−06	2.43E−05	5.03E−06	−13.97	−30.08
s526	129	21	2.70E−05	7.19E−06	2.18E−05	5.03E−06	−19.33	−30.08
styr	385	5	2.41E−05	1.20E−06	2.50E−05	7.25E−07	3.71	−39.49
s382	100	21	2.71E−05	7.19E−06	2.05E−05	1.50E−06	−24.40	−79.21
s510	152	6	8.56E−06	9.59E−07	8.53E−06	8.54E−07	−0.39	−10.98
s400	101	21	2.82E−05	7.19E−06	2.09E−05	1.50E−06	−25.76	−79.21
s641	163	19	1.01E−05	1.98E−06	9.82E−06	1.84E−06	−3.10	−6.92
s832	187	5	9.15E−06	8.27E−07	9.20E−06	2.78E−07	0.52	−66.34
s713	157	19	1.01E−05	1.98E−06	9.76E−06	1.84E−06	−2.95	−6.92
s953n	293	6	1.63E−05	1.07E−06	1.31E−05	3.64E−07	−19.22	−65.85
scf	670	7	1.77E−05	8.83E−07	1.81E−05	4.68E−07	2.40	−47.04
s1238	389	18	3.92E−05	3.45E−06	3.87E−05	2.56E−06	−1.47	−25.75
s1423	388	74	5.39E−05	1.27E−05	5.13E−05	1.13E−05	−4.79	−11.02
s1494	403	6	2.06E−05	1.51E−06	1.83E−05	5.17E−07	−11.14	−65.86
s1488	406	6	2.11E−05	1.51E−06	1.90E−05	5.17E−07	−9.82	−65.86
s5378	858	164	8.53E−05	1.71E−05	7.18E−05	1.18E−05	−15.77	−30.99
s13207	1,914	669	2.52E−04	7.64E−05	1.34E−04	3.52E−05	−46.84	−53.93
s15850	2,246	597	3.93E-04	1.15E−04	2.65E−04	5.84E−05	−32.65	−49.00
Average							−12.03	−39.56

	Com.	Seq.	Original		Clock-gated and rewired			
			D	C	D	C	δD (%)	δC (%)
s27	10	3	4.01E−06	9.59E−07	4.03E−06	7.48E−07	0.45	−21.95
s526ns	81	19	2.47E−05	6.51E−06	2.03E−05	4.34E−06	−17.80	−33.25
sse	91	4	1.15E−05	1.07E−06	1.07E−05	7.15E−07	−7.02	−32.91
s713s	123	14	7.88E−06	1.46E−06	7.32E−06	1.39E−06	−7.08	−4.73
s641-r	157	18	1.02E−05	1.88E−06	9.34E−06	1.60E−06	−8.42	−14.66
s641-retime	157	18	1.02E−05	1.88E−06	9.06E−06	1.60E−06	−10.78	−14.66
s386	85	6	1.03E−05	1.60E−06	7.28E−06	6.24E−07	−29.49	−60.98
s444	98	21	2.66E−05	7.19E−06	1.53E−05	1.38E−06	−42.31	−80.84
s526n	126	21	2.83E−05	7.19E−06	2.28E−05	5.03E−06	−19.37	−30.08
s526	129	21	2.70E−05	7.19E−06	2.14E−05	5.03E−06	−20.74	−30.08
styr	385	5	2.41E−05	1.20E−06	2.21E−05	7.25E−07	−8.11	−39.49

(*continued overleaf*)

Table 5.24 (*continued*)

	Com.	Seq.	Original		Clock-gated			
			D	C	D	C	δD (%)	δC (%)
s382	100	21	2.71E−05	7.19E−06	1.91E−05	1.50E−06	−29.56	−79.21
s510	152	6	8.56E−06	9.59E−07	8.47E−06	8.54E−07	−1.14	−10.98
s400	101	21	2.82E−05	7.19E−06	1.94E−05	1.50E−06	−31.19	−79.21
s641	163	19	1.01E−05	1.98E−06	9.11E−06	1.84E−06	−10.06	−6.92
s832	187	5	9.15E−06	8.27E−07	8.47E−06	2.78E−07	−7.38	−66.34
s713	157	19	1.01E−05	1.98E−06	9.24E−06	1.84E−06	−8.19	−6.92
s953n	293	6	1.63E−05	1.07E−06	1.19E−05	3.64E−07	−26.72	−65.85
scf	670	7	1.77E−05	8.83E−07	1.65E−05	4.68E−07	−6.84	−47.04
s1238	389	18	3.92E−05	3.45E−06	3.74E−05	2.76E−06	−4.71	−20.19
s1423	388	74	5.39E−05	1.27E−05	5.12E−05	1.13E−05	−5.07	−11.02
s1494	403	6	2.06E−05	1.51E−06	1.82E−05	5.17E−07	−11.85	−65.86
s1488	406	6	2.11E−05	1.51E−06	1.83E−05	5.17E−07	−13.10	−65.86
s5378	858	164	8.53E−05	1.71E−05	6.98E−05	1.18E−05	−18.15	−30.99
s13207	1,914	669	2.52E−04	7.64E−05	9.58E−05	2.28E−05	−61.99	−70.16
s15850	2,246	597	3.93E−04	1.15E−04	2.47E−04	5.32E−05	−37.06	−53.54
Average							−17.06	−40.14

Com. = Count of combinational cells.
Seq. = Count of sequential cells.
D = Total dynamic power.
C = Clock power.
δD = Change in total dynamic power (%).
δC = Change in clock power (%).

Table 5.22 shows the statistics of cell area required by the benchmarks. Table 5.23 list the power consumption of the original circuits, the clock gated circuits without rewiring optimization, and the clock gated circuits with rewiring optimization, respectively. The unit of power is watts [W]. The total dynamic power is the sum of the internal cell power and the net switching power. As we can see from the results, clock power is the major source of the switching power consumption.

The experimental results show that, compared to the original circuits, a very significant net reduction of 39.56% on clock power and an average saving of 12% on total dynamic power could be obtained. There was only an average 6% increase in the total cell area. By applying rewiring optimization, the total dynamic power saving could be further improved to 17%. Surprisingly, the cell area required could be reduced by 1%.

References

P. Babighian, L. Benini, and E. Macii. A scalable algorithm for RTL insertion of gated clocks based on ODCs computation. *IEEE Transactions on Computer-Aided Design of Integrated Circuits and Systems*, 24(1):29–42, 2005. ISSN: 0278-0070. doi: 10.1109/TCAD.2004.839489(410) 24.

P. Babighian, G. Kamhi, and M. Vardi. PowerQuest: trace driven data mining for power optimization. *Design, Automation & Test in Europe Conference & Exhibition, 2007. DATE '07*, pages 1–6, April 2007. doi: 10.1109/DATE.2007.364437.

L. Benini, G. De Micheli, E. Macii, M. Poncino, and R. Scarsi. Symbolic synthesis of clock-gating logic for power optimization of synchronous controllers. *ACM Transactions on Design Automation of Electronic Systems*, 4(4):351–375, 1999. ISSN: 1084-4309. doi: 10.1145/323480.323482.

L. Benini, P. Siegel, and G. De Micheli. Automatic synthesis of gated clocks for power reduction in sequential circuits. In *IEEE Design and Test of Computers*, pages 32–40, 1994.

Berkeley Logic Synthesis and Verification Group. ABC: A system for sequential synthesis and verification, release 70911. URL http://www.eecs.berkeley.edu/alanmi/abc.

R. K. Brayton. Compatible observability don't cares revisited. In *Computer Aided Design, 2001. ICCAD 2001. IEEE/ACM International Conference on*, pages 618–623, 2001. doi: 10.1109/ICCAD.2001.968725.

S. C. Chang, K. T. Cheng, N. S. Woo, and M. Marek-Sadowska. Post-layout logic restructuring using alternative wires. *IEEE Transactions on Computer-Aided Design of Integrated Circuits and Systems*, 6:1096–1106, 1999.

S. C. Chang and M. Marek-Sadowska. Perturb and simplify: multilevel Boolean network optimizer. *IEEE Transactions on Computer-Aided Design of Integrated Circuits and Systems*, 15(12):1494–1504, 1996.

Y.-P. Chen, J.-W. Fang, and Y.-W. Chang. ECO timing optimization using spare cells. pages 530–535, November 2007.

Y.-C. Chen and C.-Y. Wang. Fast detection of node mergers using logic implications. In *Computer Aided Design, 2009. ICCAD 2009. IEEE/ACM International Conference on*, 2009.

Y.-C. Chen and C.-Y. Wang. Node addition and removal in the presence of don't cares. In *Design Automation Conference (DAC), 2010 47th ACM/IEEE*, pages 505–510, 2010.

J. Cong, J. Y. Lin, and W. Long. A new enhanced SPFD rewiring algorithm. In *Proceedings of the 2002 IEEE/ACM International Conference on Computer-Aided Design*, ICCAD '02, pages 672–678, New York, 2002. ACM. ISBN: 0-7803-7607-2. doi: 10.1145/774572.774671.

J. Cong and K. Minkovich. Optimality study of logic synthesis for LUT-based FPGAs. *IEEE Transactions on Computer-Aided Design of Integrated Circuits and Systems*, 26(2):230–239, 2007.

M. Damiani and G. De Micheli. Observability don't care sets and Boolean relations. *Computer-Aided Design, 1990. ICCAD-90. Digest of Technical Papers., 1990 IEEE International Conference on*, pages 502–505, November 1990. doi: 10.1109/ICCAD.1990.129965.

M. Donno, E. Macii, and L. Mazzoni. Power-aware clock tree planning. In *ISPD '04: Proceedings of the 2004 International Symposium on Physical Design*, pages 138–147, New York, 2004. ACM. ISBN: 1-58113-817-2. doi: 10.1145/981066.981097.

R. Fraer, G. Kamhi, and M. K. Mhameed. A new paradigm for synthesis and propagation of clock gating conditions. In *DAC '08: Proceedings of the 45th Annual Conference on Design Automation*, pages 658–663, New York, 2008. ACM. ISBN: 978-1-60558-115-6. doi: 10.1145/1391469.1391638.

K.-H. Ho, Y.-P. Chen, J.-W. Fang, and Y.-W. Chang. ECO timing optimization using spare cells and technology remapping. *IEEE Transactions on Computer-Aided Design of Integrated Circuits and Systems*, 29(5):697–710, 2010.

A. Hurst. Fast synthesis of clock gates from existing logic. In *International Workshop on Logic Synthesis (IWLS) 2007*, 2007.

A. Kuehlmann. Dynamic transition relation simplification for bounded property checking. In *Computer Aided Design, 2004. ICCAD 2004. IEEE/ACM International Conference on*, pages 50–57, 2004.

T.-K. Lam, S. Yang, W.-C. Tang, and Y.-L. Wu. Logic synthesis for low power using clock gating and rewiring. In *Proceedings of the 20th Symposium on Great Lakes Symposium on VLSI*, GLSVLSI '10, pages 179–184, New York, 2010. ACM. ISBN: 978-1-4503-0012-4. doi: 10.1145/1785481.1785527.

H. Li, S. Bhunia, Y. Chen, T. N. Vijaykumar, and K. Roy. Deterministic clock gating for microprocessor power reduction. *High-Performance Computer Architecture, 2003. HPCA-9 2003. Proceedings. The Ninth International Symposium on*, pages 113–122, February 2003. ISSN: 1530-0897. doi: 10.1109/HPCA.2003.1183529.

A. Ling, D. P. Singh, and S. D. Brown. FPGA technology mapping: a study of optimality. In *28th ACM/IEEE Design Automation Conference*, pages 427–432. 2005.

V. Manohararajah, S. D. Brown, and Z. G. Vranesic. Heuristics for area minimization in LUT-based FPGA technology mapping. *IEEE Transactions on Computer-Aided Design of Integrated Circuits and Systems*, 25(11):2331–2340, 2006.

A. Marquardt, V. Betz, and J. Rose. Timing-driven placement for FPGAs. In *Proceedings of the ACM/SIGDA International Symposium on FPGAs*, pages 203–213. 2000.

L. McMurchie and C. Ebeling. Pathfinder: a negotiation-based performance-driven router for FPGAs. In *Proceedings of the Field-Programmable Gate Arrays (FPGA'95)*, pages 111–117, 1995.

A. Mishchenko, S. Chatterjee, and R. Brayton. DAG-aware AIG rewriting a fresh look at combinational logic synthesis. In *DAC '06: Proceedings of the 43rd Annual Design Automation Conference*, pages 532–535, New York, 2006. ACM. ISBN: 1-59593-381-6. doi: 10.1145/1146909.1147048.

A. Mishchenko, S. Chatterjee, and R. Brayton. Improvements to technology mapping for LUT-based FPGAs. *IEEE Transactions Computer-Aided Design of Integrated Circuits and Systems*, 26(2):240–253, 2007 10.1109/TCAD.2006.887925.

S. M. Plaza, K.-H. Chang, I. L. Markov, and V. Bertacco. Node mergers in the presence of don't cares. In *ASP-DAC '07: Proceedings of the 2007 Asia and South Pacific Design Automation Conference*, pages 414–419, Washington, DC, 2007. IEEE Computer Society. ISBN: 1-4244-0629-3. doi: 10.1109/ASPDAC.2007.358021.

J. A. Roy, S. N. Adya, D. A. Papa, and I. L. Markov. Min-cut floorplacement. *IEEE Transactions on Computer-Aided Design of Integrated Circuits and Systems*, 25(7):1313–1326, 2006.

N. Saluja and S. P. Khatri. A robust algorithm for approximate compatible observability don't care (CODC) computation. In *DAC '04: Proceedings of the 41st Annual Design Automation Conference*, pages 422–427, New York, 2004. ACM. ISBN: 1-58113-828-8. doi: 10.1145/996566.996688.

N. Vemuri, P. Kalla, and R. Tessier. BDD-based logic synthesis for LUT-based FPGAs. *ACM Transactions on Design Automation of Electronic Systems*, 7(4):501–525, 2002.

Y.-L. Wu and M. Marek-Sadowska. Orthogonal greedy coupling - a new optimization approach to 2-D FPGA routing. In *DAC '95: Proceedings of the 32nd ACM/IEEE Design Automation Conference*, 1995.

X. Yang, T.-K. Lam, and Y.-L. Wu. ECR: a low complexity generalized error cancellation rewiring scheme. In *DAC '10: Proceedings of the 47th Design Automation Conference*, pages 511–516, New York, 2010. ACM. ISBN: 978-1-4503-0002-5. doi: 10.1145/1837274.1837400.

L. Zhou, W. C. Tang, and Y. L. Wu. Fast placement-intact logic perturbation targeting for FPGA performance improvement. In *3rd Southern Conference on Programmable Logic*, pages 63–68, 2007a.

C. L. Zhou, W.-C. Tang, W.-H. Lo, and Y.-L. Wu. How much can logic perturbation help from netlist to final routing for FPGAs. In *44th ACM/IEEE Design Automation Conference*, pages 922–927, 2007b.

Q. Zhu, N. Kitchen, A. Kuehlmann, and A. Sangiovanni-Vincentelli. SAT sweeping with local observability don't-cares. In *DAC '06: Proceedings of the 43rd Annual Design Automation Conference*, pages 229–234, New York, 2006. ACM. ISBN: 1-59593-381-6. doi: 10.1145/1146909.1146970.

6

Summary

We have discussed various rewiring techniques, including their concepts and their unique ways of replacing some wires with other additional wires in the circuit. Their applications on some selected electronic design automation (EDA) areas have also been introduced. It can been seen that certain rewiring techniques are more suitable and effective to achieve certain goals.

What we have not discussed in detail in this book is the strategy of applying rewiring. For some specific optimization problems, it is quite easy to formulate some efficient heuristics to select target wires and alternative wires, and the orders of wire replacements. In general, however, deriving such effective heuristics requires a deep understanding of the optimization problem and the properties of rewiring techniques, and therefore is not an easy task. This is especially true when rewiring, which, being a logical operation, is applied to optimize the circuit in the phase of physical design.

Bridging logical and physical design processes (front-end and back-end) is in fact another area on which researchers and engineers want to spend effort. We have demonstrated the effectiveness of considering logical transformations in physical design optimizations. Since rewiring is wire-oriented and wire delay is the dominant factor determining the timing properties of the circuit, it is well suited to fixing timing closure.

We hope readers will find this book informative and useful to their studies and areas of work.

Boolean Circuit Rewiring: Bridging Logical and Physical Designs, First Edition.
Tak-Kei Lam, Wai-Chung Tang, Xing Wei, Yi Diao and David Yu-Liang Wu.

Index

Boolean Circuit Rewiring: Bridging Logical and Physical Designs, First Edition.
Tak-Kei Lam, Wai-Chung Tang, Xing Wei, Yi Diao and David Yu-Liang Wu.